Information Theory

Part I: An Introduction to the Fundamental Concepts

Information Theory

Part I: An Introduction to the Fundamental Concepts

Arieh Ben-Naim

The Hebrew University of Jerusalem, Israel

NEW JERSEY • LONDON • SINGAPORE • BEIJING • SHANGHAI • HONG KONG • TAIPEI • CHENNAI • TOKYO

Published by

World Scientific Publishing Co. Pte. Ltd.
5 Toh Tuck Link, Singapore 596224
USA office: 27 Warren Street, Suite 401-402, Hackensack, NJ 07601
UK office: 57 Shelton Street, Covent Garden, London WC2H 9HE

Library of Congress Cataloging-in-Publication Data
Names: Ben-Naim, Arieh, 1934–
Title: Information theory / Arieh Ben-Naim, The Hebrew University of Jerusalem, Israel.
Description: New Jersey : World Scientific, 2017– | Includes bibliographical
references and index. Contents: part 1. An introduction to the fundamental concepts
Identifiers: LCCN 2017015276| ISBN 9789813208827 (hardcover : alk. paper : pt. 1) |
ISBN 9813208821 (hardcover : alk. paper : pt. 1) |
ISBN 9789813208834 (pbk. : alk. paper : pt. 1) |
ISBN 981320883X (pbk. : alk. paper : pt. 1)
Subjects: LCSH: Information theory. | Entropy (Information theory)
Classification: LCC Q370 .B46 2017 | DDC 003/.54--dc23
LC record available at https://lccn.loc.gov/2017015276

British Library Cataloguing-in-Publication Data
A catalogue record for this book is available from the British Library.

Printed in Singapore

This book is dedicated to all those who confuse the concepts of
Information with Shannon's Measure of Information
and Shannon's Measure of Information with Entropy.

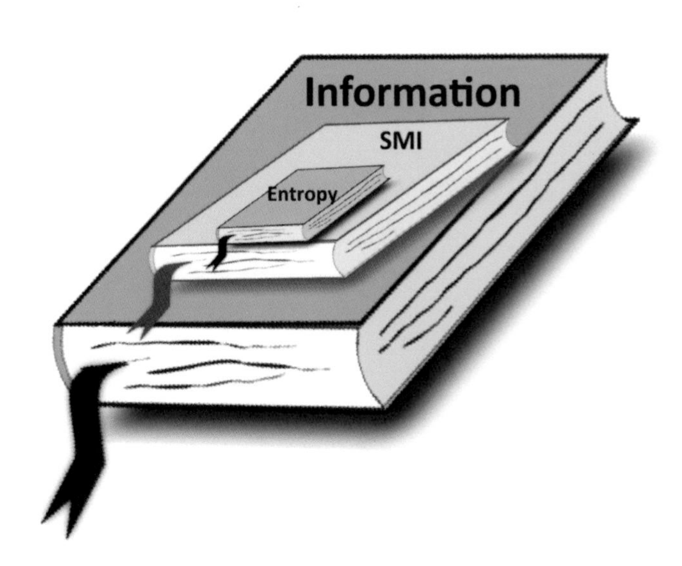

If there is no knowledge there is no understanding; אִם אֵין בִּינָה, אֵין דָּעַת. אִם אֵין דַּעַת, אֵין בִּינָה.

if there is no understanding there is no knowledge. אִם אֵין קֶמַח, אֵין תּוֹרָה. אִם אֵין תּוֹרָה, אֵין קֶמַח.

פרקי אבות, פרק ג, משנה יז

Contents

Preface

This book has been written for those who are or may be using the concepts of information theory in studying problems which are considered outside the realm of information theory.

There are many excellent textbooks on information theory which are addressed to professionals or professionals-to-be in the field. These books are highly mathematical, and may not be appealing to readers who are not interested in all the subtle mathematical details but still want to use the concepts of information theory within their field of research.

On the other hand, there are many popular science books which are easy to read but are not helpful to those who are interested in the application of information theory. In addition, most of these books contain qualitative, sometimes misleading information on information theory.

Thus, the aim of this book is to provide the reader with an introduction to the fundamental concepts of information theory, using simple, clear, and reader-friendly language.

To this end, I have used minimal mathematics which is essential for understanding the basis of information theory, relegating less essential mathematical details to either appendices or notes. For more comprehensive mathematical details, the reader is referred to the relevant literature.

In order to read and understand this book, one needs to know the basics of calculus and some matrix algebra. Throughout it, examples are worked out in detail, and a few exercises are suggested so that the reader can self-test his or her understanding of the concepts.

Having taught thermodynamics, information theory, and probability for over 20 years, I am convinced that:

(1) To understand entropy and the second law, it is indispensable to understand the basics of information theory.
(2) To understand information theory, it is indispensable to understand the basics of probability theory.

With this conviction I constructed the book in three parts; the main part consists of four chapters (1–4), which are bookended by Chap. 0, on probability, and Chap. 5, on entropy and the second law.

I hope that the readers of this book will find the chapters on information theory useful in whatever field they are interested in. I also hope that they will be convinced that it is of the utmost importance to make a clear-cut distinction between entropy and Shannon's measure of information (SMI). Although the former is defined and interpreted in terms of the latter, these two concepts should be distinguished. SMI is extremely general, and can be applied to any probability distribution. On the other hand, entropy is defined only on a tiny set of distributions, specifically those distributions pertaining to macroscopic systems at equilibrium.

As always, I would be grateful to receive any comments or suggestions from readers.

Arieh Ben-Naim

Department of Physical Chemistry
The Hebrew University of Jerusalem
Jerusalem, Israel
Email: ariehbook@gmail.com
URL: ariehbennaim.com

Acknowledgments

I am grateful to the many readers of my previous books who sent me useful suggestions, which were implemented in writing this book.

I am also grateful to Diego Casadei, Claude Dufour, Robert Engel, Jose Angel Sordo Gonzalo, Douglas Hemmick, Shannon Hunter, Bernard Lavenda, Lester Lipsky, Thierry Lorho, and Mike Rainbolt, who read parts or the whole manuscript and offered useful comments and suggestions. I am grateful, too, to Alex Vaisman for the figure he drew, which I have used for the cover design of the book.

As always, I am most grateful for the help of my wife, Ruby. Without her assistance I would never have completed this book.

List of Abbreviations

1D	one-dimensional
CI	conditional mutual information
IT	information theory
KSA	Kirkwood superposition approximation
MI	mutual information
PMF	potential of mean force
20Q	twenty questions
rv	random variable
SMI	Shannon's measure of information
TI	total mutual information

0

Elements of Probability Theory

0.1 Introduction

Probability theory is a branch of mathematics. Its uses are in all fields, from physics and chemistry to biology and sociology, and to economics and psychology; in short, it is employed everywhere and anytime in our lives. This chapter presents the minimum amount of probability theory that one needs to understand information theory (IT). The reader who is familiar with the theory of probability can skip this chapter, and start at Chapter 1.

It is an experimental fact that in many seemingly random events, some regularities are observed. For instance, in throwing a die one cannot predict any single outcome. However, if one throws the die many times, then one will observe that the relative frequency of occurrence of, say, the outcome "3" is about 1/6. We also say that the *probability* of occurrence of the outcome "3" is 1/6.

In the previous sentence we have used the term "probability" without defining it. In fact, there have been several attempts to *define* this term. As it turns out, each definition has its limitations. But, more importantly, each definition uses the concept of probability in its very definition, i.e. all definitions are circular. Nowadays, the mathematical theory of probability has an axiomatic basis, much like Euclidean geometry or any other branch of mathematics. Probability is considered to be a primitive concept that cannot be defined in terms of more primitive concepts, much as a point or a line in geometry is considered to be a primitive concept. Thus, even without having a proper definition of the term, probability theory is a fascinating and extremely useful concept.

Probability theory was developed mainly in the 17th century, by the mathematicians Pierre de Fermat (1601–1665), Blaise Pascal (1623–1662), Christiaan Huygens (1629–1695), Jacob Bernoulli (1654–1705), and many others. The chief motivation for developing the mathematical theory of probability was to answer various questions regarding games of chance. For literature on the history of probability and some textbooks, see the bibliography.

In this chapter, we will start with the axiomatic approach to probability theory. We will discuss a few methods of calculating probabilities and a few theorems that are useful in IT.

0.2 The Axiomatic Approach

The axiomatic approach to probability was developed mainly by Andrey Kolmogorov (1903–1987), in 1930's. It consists of three elements, denoted as $\{\mathbf{\Omega}, \mathbf{F}, \mathbf{P}\}$, which together define the probability *space*.

0.2.1 *The Sample Space, Ω*

This is the set of all possible outcomes of a specific experiment (sometimes referred to as a trial).

Examples

The sample space of all possible outcomes of tossing a coin consists of two elements: $\mathbf{\Omega} = \{\mathrm{H}, \mathrm{T}\}$, where H stands for "head" and

T "tail." The sample space of throwing a die consists of six possible outcomes: $\mathbf{\Omega} = \{1, 2, 3, 4, 5, 6\}$. These are called simple events or "points" in the sample space. In most cases, simple events are equally probable, in which case they are referred to as *elementary events*.

Clearly, we cannot write down the sample space for every experiment. Some consist of an infinite number of elements (for example, the point hit when one is shooting an arrow at a board), and some cannot even be described (for example, what the world will look like next year). We will be interested only in simple sample spaces where the counting of the elementary (or simple) events is straightforward.

0.2.2 *The Field of Events, F*

A compound event, or simply an *event*, is defined as a union (or a sum) of elementary events. Examples of events are:

(a) The result "even" of tossing a die consists of the elementary events $\{2, 4, 6\}$. This means that 2 or 4 or 6 has occurred, or will occur in the experiment of tossing a die.
(b) The result "larger than or equal to 5" of tossing a die consists of the elementary events $\{5, 6\}$, i.e. either 5 or 6 has occurred.

In mathematical terms, $\boldsymbol{F}$ consists of all partial sets of the sample space $\mathbf{\Omega}$. Note that the event $\mathbf{\Omega}$ (i.e. the complete set of elementary events) itself belongs to $\boldsymbol{F}$. The empty event denoted as $\emptyset$ also belongs to $\boldsymbol{F}$.

We will mainly discuss finite sample spaces. We will also apply some of the results to infinite or even continuous spaces by using arguments of analogy. A more rigorous treatment requires the tools of measure theory.

For a finite sample space, each partial set of $\mathbf{\Omega}$ is an event. If there are n elementary events in $\mathbf{\Omega}$, then the total number of events in $\boldsymbol{F}$ is 2^n. The proof is obtained by straightforward counting:[a]

$$\begin{aligned}
&\binom{n}{0}, \text{ one event denoted as } \phi \text{ (the impossible or empty event);}\\
&n = \binom{n}{1}, \text{ simple (or elementary) events;}\\
&\binom{n}{2}, \text{ events consisting of two simple events;}\\
&\binom{n}{3}, \text{ events consisting of three simple events;}\\
&\vdots\\
&\binom{n}{n}, \text{ one event consisting of all the space } \boldsymbol{\Omega} \text{ (the certain event).}
\end{aligned} \tag{0.1}$$

Altogether, we have 2^n events, i.e.

$$\binom{n}{0} + \binom{n}{1} + \binom{n}{2} + \cdots + \binom{n}{n} = (1+1)^n = 2^n. \tag{0.2}$$

Exercise 0.1

Check that for $n = 3$, $\sum_{i=0}^{3} \binom{3}{i} = 2^3 = 8$.

We have used Newton's binomial theorem in performing the summation in Eq. (0.2). The general form of the theorem is

$$(x+y)^n = \sum_{i=0}^{n} \binom{n}{i} x^i y^{n-i}. \tag{0.3}$$

---

[a]The symbol $\binom{n}{m}$ means $\frac{n}{(n-m)!m!}$. By definition $0! = 1$, and hence $\binom{n}{0} = 1$.

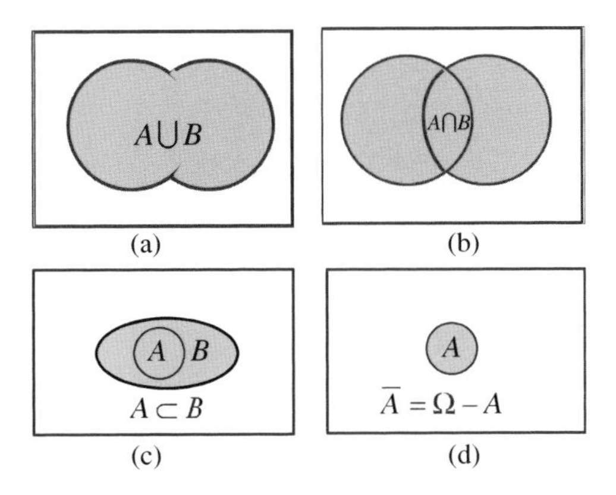

Fig. 0.1. (a) Union (or sum) of events; (b) intersection (or product) of events; (c) A is included in B; and (d) $\bar{A}$ is complementary (or not A) to A.

At this stage, we introduce some notations regarding operations between events:

The event $\boldsymbol{A} \cup \boldsymbol{B}$ (or $\boldsymbol{A}+\boldsymbol{B}$) is called the *union* (or the sum) of the *two events*. This is the event "either $\boldsymbol{A}$ or $\boldsymbol{B}$ has occurred."

The event $\boldsymbol{A} \cap \boldsymbol{B}$ (or $\boldsymbol{A} \cdot \boldsymbol{B}$) is called the *intersection* (or the product) of the *two events*. This is the event "both $\boldsymbol{A}$ and $\boldsymbol{B}$ have occurred."

The complementary event denoted as $\bar{A}$ (or $\boldsymbol{\Omega}$–$\boldsymbol{A}$) is the event "$\boldsymbol{A}$ did not occur."

The notation $\boldsymbol{A} \subset \boldsymbol{B}$ means that $\boldsymbol{A}$ is *partial* to $\boldsymbol{B}$, or $\boldsymbol{A}$ is *included* in the event $\boldsymbol{B}$, i.e. the occurrence of $\boldsymbol{A}$ implies the occurrence of $\boldsymbol{B}$.

These relationships between events are described in Fig. 0.1.

0.2.3 *The Probability Function P*

For each event A belonging to F, we assign a number P, called the *probability* of the event A. This number fulfills the following properties:

$$a : P(\boldsymbol{\Omega}) = 1; \tag{0.4}$$

we will use either P or p, and sometimes Pr, to denote probability;

$$b : 0 \le P(A) \le 1; \tag{0.5}$$

c : If $\boldsymbol{A}$ and $\boldsymbol{B}$ are disjoint events (or mutually exclusive), then

$$P(\boldsymbol{A} \cup \boldsymbol{B}) = P(\boldsymbol{A}) + P(\boldsymbol{B}). \tag{0.6}$$

The first two conditions define the range of numbers for the probability function. The first condition simply means that the event $\boldsymbol{\Omega}$ has the largest value of the probability. By definition, we assume that one outcome has occurred, or will occur. Hence, the event $\boldsymbol{\Omega}$ is also referred to as the *certain event*, and is assigned the value of 1. The impossible event is denoted as ϕ assigned the number 0, i.e. $P(\phi) = 0$.

The third condition is intuitively clear. Two events $\boldsymbol{A}$ and $\boldsymbol{B}$ are said to be *disjoint*, or *mutually exclusive*, when the occurrence of one event excludes the possibility of the occurrence of the other. In mathematical terms, we say that the *intersection* of the two events $(\boldsymbol{A} \cap \boldsymbol{B})$ is empty, i.e. there is no elementary event that is common to $\boldsymbol{A}$ and $\boldsymbol{B}$. For example, the two events

$$\boldsymbol{A} = \{\text{the outcome of throwing a die is even}\},$$

$$\boldsymbol{B} = \{\text{the outcome of throwing a die is odd}\}.$$

Clearly, the events $\boldsymbol{A}$ and $\boldsymbol{B}$ are disjoint; the occurrence of one *excludes* the occurrence of the other. Next, we define the event

$$\boldsymbol{C} = \{\text{the outcome of throwing a die is larger than or equal to 5}\}.$$

Clearly, $\boldsymbol{A}$ and $\boldsymbol{C}$, or $\boldsymbol{B}$ and $\boldsymbol{C}$, are not disjoint. $\boldsymbol{A}$ and $\boldsymbol{C}$ contain the elementary event 6. $\boldsymbol{B}$ and $\boldsymbol{C}$ contain the elementary event 5.

The events "greater than or equal to 4" and "smaller than or equal to 2" are clearly disjoint or mutually exclusive events.

If the two events are not disjoint, say "greater than or equal to 4" and "even," the rule (c), Eq. (0.6) should be modified into

$$d : P(\boldsymbol{A} \cup \boldsymbol{B}) = P(\boldsymbol{A}) + P(\boldsymbol{B}) - P(\boldsymbol{A} \cap \boldsymbol{B}). \tag{0.7}$$

A very useful way of demonstrating the concept of probability and the sum rule is the Venn diagram (Fig. 0.1).

Suppose that we throw a dart at a rectangular board having a total area of $\boldsymbol{\Omega}$. We assume that the dart hits some point within the board. We now draw a circle within the board (Fig. 0.2), and ask what the probability of hitting the area within this circle is. We assume, by common sense, that the probability of the event "hitting inside the circle" is equal to the ratio of the area of the circle to the area of the entire board.[b]

Two regions drawn on the board are said to be disjoint if there is no overlap between them. It is clear that the probability of hitting either one region or the other is the ratio of the area of the two regions to the area of the whole board.

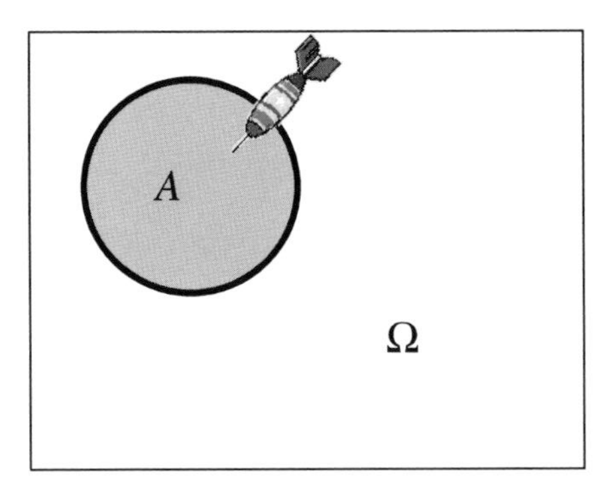

Fig. 0.2. The probability of hitting the region A is equal to the ratio of the area of the region A to the total area of the rectangle, Ω.

[b]We exclude the probability of hitting a specific point or a line — these have zero probability.

This leads directly to the sum rules stated in the axioms above. The probability of hitting either of the regions is the sum of the probabilities of hitting each of the regions.

This sum rule (0.6) does not hold when the two regions overlap, i.e. when there are points on the board that belong to both regions. It is quite clear that the probability of hitting either of the regions ***A*** and ***B*** is, in this case, the sum of the probabilities of hitting each of the regions minus the probability of hitting the overlapping region.

On this relatively simple axiomatic foundation, the whole edifice of the mathematical theory of probability has been erected. It is not only extremely useful but is also an essential tool in all the sciences and beyond.

In the axiomatic structure of the theory of probability, the probabilities are said to be *assigned* to each event. These probabilities must fulfill the four conditions: a, b, c, and d. The theory does not *define* probability, or provide a method of calculating or measuring these probabilities. In fact, there is no way of calculating probabilities for any general event. It is still a quantity that measures our degree of belief in the occurrence of certain events. As such, it is a highly subjective quantity. However, for some simple experiments, say tossing a coin or throwing a die, we have some very useful methods of calculating probabilities. They have their limitations and they apply to "ideal" cases, yet these probabilities turn out to be extremely useful. What is more important, since these probabilities are based on common sense, is that we should *all* agree that they are the "correct" probabilities, i.e. these probabilities turn from subjective quantities into objective quantities. They "belong" to the events as much as the mass belongs to a piece of matter. We will describe two very useful "definitions" that have been suggested and commonly used for this concept.

In everyday life, we use the term "probability" or "chances" to express our extent of belief in some proposition. These probabilities have nothing to do with the theory of probability.

0.3 The Classical "Definition" of Probability

This is sometimes referred to as the *a priori* definition. Let $N(\text{total})$ be the *total* number of outcomes of an experiment (for example, for throwing a die $N(\text{total})$ is 6). We denote by $N(\text{event})$, the number of outcomes (i.e. elementary events) that are included in the event in which we are interested. The probability of the event in which we are interested is *defined* as the ratio

$$\mathrm{P}(\text{event}) = \frac{N(\text{event})}{N(\text{total})}. \tag{0.8}$$

Care must be taken in applying Eq. (0.8) as a *definition* of probability. First, not every event can be "decomposed" into elementary events (for example, the event "tomorrow it will start raining at ten o'clock"). But, more importantly, the above formula presumes that each of the elementary events has the same likelihood of occurrence. In other words, each elementary event is presumed to have the same *probability*, such as 1/6 in the case of a die. This is the reason why the classical "definition" cannot be used as a *bona fide definition* of the concept of probability; it is a circular definition. Nevertheless, this "definition" (or, rather, the method of calculating probabilities) is extremely useful.

0.4 The Relative Frequency "Definition" of Probability

This definition is referred to as the *a posteriori* or experimental definition, since it is based on actual counting of the relative frequency of occurrence of events.

The simplest example would be tossing a coin. There are two possible outcomes: head (H) or tail (T). We exclude the rare events where the coin will fall exactly perpendicularly to the floor or break into pieces during the experiment.

We toss the coin N times. The frequency of occurrence of heads is recorded. This is a well-defined and feasible experiment. If $N(H)$ is the number of heads that have occurred in N(total) trials, then the frequency of occurrence of heads is N(H)/N(total). The *probability* of occurrence of the event H is *defined* as the limit of the frequency N(H)/N(total) when N(total) tends to infinity. The frequency definition is thus

$$P(H) = \lim_{N(\text{total}) \to \infty} \frac{N(H)}{N(\text{total})}. \tag{0.9}$$

Clearly, such a definition is not practical: first, because we cannot perform an infinite number of trials; second, even if we could, who could guarantee that such a limit exists? We believe that such a limit exists and that it is unique; but, in fact, we cannot prove that.

In practice, we use this definition for very large number N. Why? Because we believe that if N is large enough and if the coin is fair, then there is a *high probability* that the relative frequency of occurrence of heads will be 1/2. We see that we have again used the *concept* of probability in the very definition of probability.

0.5 Independent Events and Conditional Probability

The concepts of dependence between events and conditional probability are central to probability theory and have many uses in the sciences.[c] Two events are said to be *independent* if the occurrence of one event has no effect on the probability of occurrence of the other. Mathematically, two events A and B are said to be independent if and

[c]Note also that until now we have discussed relations between events that are relations between sets of outcomes, as in set theory. Dependence or independence between events is a relation that is unique to probability theory — and has no counterpart in set theory.

only if

$$P(A \cdot B) = P(A)P(B). \tag{0.10}$$

For example, if two people who are far apart from each other throw a fair die each, the outcomes of the two dice are *independent* in the sense that the occurrence of, say, "5" on one die does not have any effect on the probability of occurrence of a result that is, say, "3" on the other. On the other hand, if the two dice are connected by an inflexible wire, the two outcomes can be dependent. Similarly, if we throw a single die consecutively, and at each throw the die is deformed or damaged, the outcomes will not be independent. Intuitively, it is clear that whenever two events are independent, the probability of the occurrence of both events — say, "5" on one die and "3" on the other — is the *product* of the two probabilities. The reason is quite simple. By tossing two dice simultaneously, we have altogether 36 possible elementary events (the first die's outcome, "i" and the second die's outcome, "j"). Each of these outcomes has an equal probability of 1/36, which is also equal to 1/6 times 1/6, i.e. the product of the probabilities of each event separately.

A fundamental concept in the theory of probability is the *conditional probability*. This is the probability of the occurrence of an event A given that an event B has occurred. We write this as $P(A|B)$ (read: the Probability of A given the occurrence of B),[d] and it is defined by

$$P(A|B) = P(A \cdot B)/P(B). \tag{0.11}$$

Clearly, if the two events are independent, then the occurrence of B has no effect on the probability of the occurrence of A. Hence, from Eqs. (0.10) and (0.11), we get

$$P(A|B) = P(A). \tag{0.12}$$

[d]This definition is valid for $P(B) \neq 0$. Note also that $P(B)$ is sometimes interpreted as the probability that the *proposition* B is true.

We define the *correlation* between the two events as[e]

$$g(A,B) = \frac{P(A \cdot B)}{P(A)P(B)}. \tag{0.13}$$

We say that the two events are *positively correlated* when $g(A,B) > 1$, i.e. the occurrence of one event enhances or increases the probability of the second event. We say that the two events are *negatively correlated* when $g(A,B) < 1$, and that they are *uncorrelated* or *indifferent* when $g(A,B) = 1$.

As an example, consider the following events:

$$A = \{\text{the outcome of throwing a die is “4”}\},$$
$$B = \{\text{the outcome of throwing a die is “even”}\}$$

(i.e. it is one of the following outcomes: 2, 4, 6),

$$C = \{\text{the outcome of throwing a die is “odd”}\} \tag{0.14}$$

(i.e. it is one of the following outcomes: 1, 3, 5).

We can calculate the two conditional probabilities

$$P(\text{of } A|\text{given } B) = \frac{1}{3} > P(\text{of } A) = \frac{1}{6}, \tag{0.15}$$

$$P(\text{of } A|\text{given } C) = 0 < P(\text{of } A) = \frac{1}{6}. \tag{0.16}$$

In the first example, the knowledge that B has occurred *increases* the probability of the occurrence of A. Without that knowledge, the probability of A is $1/6$ (one out of six possibilities). Given the occurrence of B, the probability of A becomes larger, i.e. $1/3$ (one out of three possibilities). But *given* that C has occurred, the probability of A becomes zero, i.e. *smaller* than the probability of A without that knowledge.

[e] In the theory of probability, correlation is normally defined for a random variable. Here, we define the correlation between two events as the ratio between the probabilities in Eq. (0.13).

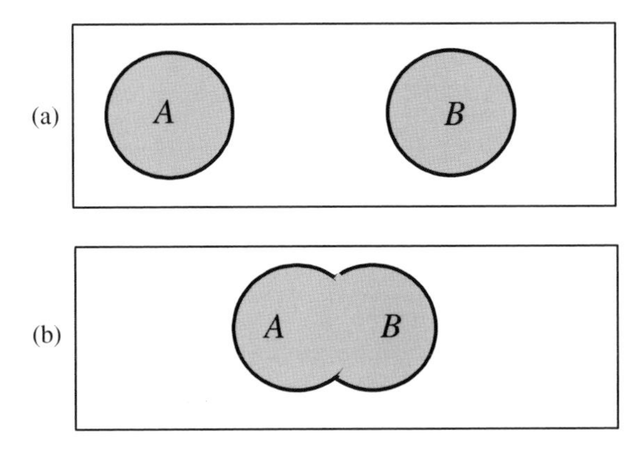

Fig. 0.3. (a) Disjoint and (b) overlapping events.

It is important to distinguish between *disjoint* events and *independent* events. Disjoint events are events that are mutually exclusive; the occurrence of one excludes the occurrence of the second. The events "even" and "5" are disjoint. In terms of Venn diagrams, two regions that are nonoverlapping are disjoint. If the dart hits one region, say A, in Fig. 0.3(a), then we know for sure that it did not hit the second region, B.

"*Disjoint*" is a property of the events themselves (i.e. the two events have no common elementary event). "*Independence*" between events is defined not in terms of the elementary events comprising the two events, but in terms of their probabilities. If the two events are *disjoint*, then they are strongly *dependent*. In the above example, we say that two events are *negatively* correlated. In other words, the conditional probability of hitting one circle, given that the dart hits the other circle, is zero (which is smaller than the "unconditional" probability of hitting one circle).

If the two regions A and B do overlap (i.e. they are not disjoint), then the two events could be either dependent or independent. In fact, the two events could either be positively or negatively correlated. The following example illustrates the relation between dependence and the extent of overlapping in Fig. 0.4.

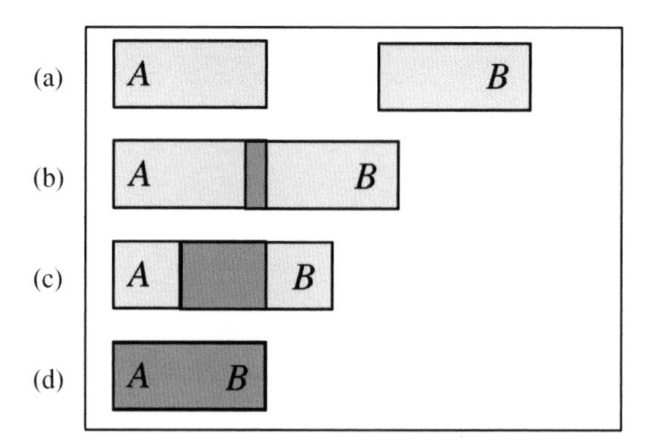

Fig. 0.4. (a) Extreme negative correlation; (b) negative correlation; (c) positive correlation; and (d) extreme positive correlation.

Consider the two rectangles $\boldsymbol{A}$ and $\boldsymbol{B}$ having the same area, $a \times b$.

Since $\boldsymbol{A}$ and $\boldsymbol{B}$ have the same area, the probability of hitting $\boldsymbol{A}$ is equal to the probability of hitting $\boldsymbol{B}$. Let us denote that by $P(\boldsymbol{A}) = P(\boldsymbol{B}) = p$, where p is the ratio of the area of $\boldsymbol{A}$ to the total area of the board — say $p = 1/10$, in this illustration.

When the two rectangles are separated, i.e. there is no overlapping area, we have

$$P(\boldsymbol{A}|\boldsymbol{B}) = 0. \tag{0.17}$$

This is the extreme *negative* correlation. Given that the dart hits $\boldsymbol{B}$, the probability that it hits $\boldsymbol{A}$ is 0 (i.e. smaller than p).

The other extreme case is when $\boldsymbol{A}$ and $\boldsymbol{B}$ coincide (fully overlap). In this case, we have

$$P(\boldsymbol{A}|\boldsymbol{B}) = 1. \tag{0.18}$$

This is the extreme *positive* correlation. Given that the dart hits $\boldsymbol{B}$, the probability that it hits $\boldsymbol{A}$ is 1 (i.e. larger than p).

When we move $\boldsymbol{B}$ toward $\boldsymbol{A}$, the correlation changes continuously from maximum *negative* correlation (nonoverlapping) to *maximum* positive correlation (total overlapping). In between, there is a point where the two events $\boldsymbol{A}$ and $\boldsymbol{B}$ are independent. Assume for simplicity that the total area of the board is unity. Then the probability of hitting A is simply the area of A. The probability of hitting B is the area of B. Let $p = P(A) = P(B)$. The conditional probability is calculated by $P(A|B) = P(A \cdot B)/P(B)$, where $P(A \cdot B)$ is the probability of hitting both A and B. Denote this probability by x. The correlation between the two events, "hitting A" and "hitting B," is defined as $P(A \cdot B)/P(A)P(B) = x/p^2$. If the overlapping area is $x = 0$, then $g = 0$ (i.e. *extreme negative correlation*). If the overlapping area is $x = p$, we have $g = p/p^2 = 1/p$ or $P(A|B) = p/p = 1$ (i.e. *extreme positive correlation*). Between these two extreme cases of $x = 0$ and $x = p$, we have one point, $x = p^2$. At this point $g = p^2/p^2 = 1$ or $P(A|B) = P(A)$; in this case there is *no correlation* between the two events, i.e. the two events are independent.

Finally, we note that the concept of independence between n events, $A_1, A_2, \ldots, A_n$, is defined by the requirement that

$$P(A_1, A_2, \ldots, A_n) = \prod_{i=1}^{n} P(A_i). \tag{0.19}$$

It should be noted that independence between the n events does not imply independence between pairs, triplets, etc. For example, independence between three events A_1, A_2, A_3 does not imply independence between pairs, and vice versa. An example is shown in Ben-Naim (2015b).

Consider the following simple and very illustrative example that was studied in great detail [Falk (1979), Ben-Naim (2015b)]. It demonstrates how we intuitively associate conditional probability with the direction of occurrence of the events in time, confusing *causality* with *conditional probabilistic argument*.

The problem is very simple. An urn contains four balls, two white and two black. The balls are well mixed and we draw one ball while blindfolded.

First, we ask: What is the probability of the event "white ball on the first draw"? The answer is immediate: 1/2. There are four equally probable outcomes; two of them are consistent with the "white ball" event, and hence the probability of the event is $2/4 = 1/2$.

Next, we ask: What is the conditional probability of drawing a white ball on a *second* draw, given that on the first draw we drew a while ball (the first ball is not returned to the urn). We write this conditional probability as $P(\text{White}_2|\text{White}_1)$. The calculation is very simple. We know that a white ball was drawn in the first trial and was not returned. After the first draw, three balls are left — two black and one white. The probability of drawing a white ball is simply $1/3$. We write this conditional probability as

$$P(\text{White}_2|\text{White}_1) = \frac{1}{3}. \tag{0.20}$$

Now, the trickier question: What is the probability that we *drew* a white ball on the *first* draw, given that the second draw was white? Symbolically, we ask for

$$P(\text{White}_1|\text{White}_2) = ? \tag{0.21}$$

This is a baffling question. How can an event in the "present" (white ball on the *second* draw) affect an event in the "past" (white ball drawn in the *first* trial)?

These questions were actually asked in a classroom. The students effortlessly answered the question about $P(\text{White}_2|\text{White}_1)$, arguing that drawing a white ball on the first draw has *caused* a change in the urn, and therefore has influenced the probability of drawing a second white ball.

However, asking about $P(\text{White}_1|\text{White}_2)$ caused an uproar in the class. Some claimed that this question is meaningless, arguing that an event in the "present" cannot affect the probability of an event in the "past." Some argued that since the event in the present cannot affect the probability of the event in the past, the answer to the question is 1/2. They were wrong. The answer is 1/3. We will revert to this problem and its solution after presenting Bayes' theorem in the next section.

0.6 Bayes' Theorem

Bayes' theorem is an extremely useful tool both for calculating probabilities and for plausible reasoning. We will start with a simple and intuitively clear theorem known as the *theorem of total probability.*

Consider n events that are pairwise disjoint so that their union (or sum) covers the entire space $\mathbf{\Omega}$. In mathematical terms, we assume that

$$(1)\ \mathbf{A}_i \cdot \mathbf{A}_j = \phi \text{ for each pair of events, } j = 1, 2, \ldots, n, \quad i \neq j; \tag{0.22}$$

$$(2)\ \mathbf{\Omega} = \sum_{i=1}^{n} \mathbf{A}_i \text{ (or } \cup_{i=1}^{n} \mathbf{A}_i). \tag{0.23}$$

Let $\mathbf{B}$ be any event. We can always write

$$\mathbf{B} = \mathbf{B} \cdot \mathbf{\Omega} = \mathbf{B} \cdot \sum_{i=1}^{n} \mathbf{A}_i = \sum_{i=1}^{n} \mathbf{B} \cdot \mathbf{A}_i. \tag{0.24}$$

The first equality is evident (any event is equal to the intersection of the event with the total space of events $\mathbf{\Omega}$). The second equality follows from the assumption (0.23). The third equality follows from the distributive law of the product of events.

From the assumption that all $\mathbf{A}_i$ are mutually exclusive, it also follows that all $\mathbf{B} \cdot \mathbf{A}_i$ are mutually exclusive. Therefore, it follows that the probability of $\mathbf{B}$ is the sum of the probabilities of all the $\mathbf{B} \cdot \mathbf{A}_i$; see Fig. 0.5.

$$P(\mathbf{B}) = P\left(\sum_{i=1}^{n} \mathbf{B} \cdot \mathbf{A}_i\right) = \sum_{i=1}^{n} P(\mathbf{B} \cdot \mathbf{A}_i) = \sum_{i} P(\mathbf{A}_i) P(\mathbf{B}|\mathbf{A}_i). \tag{0.25}$$

The last equality follows from the definition of the conditional probabilities.

The theorem (0.25) is very simple and intuitively clear. We have started with n events that cover the entire space of events $\mathbf{\Omega}$.

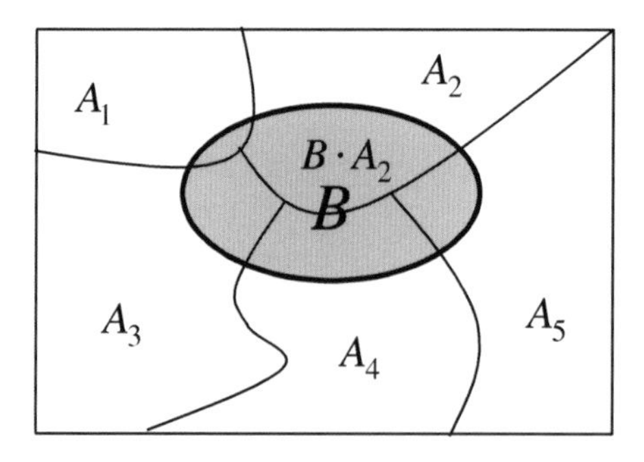

Fig. 0.5. A region $\boldsymbol{B}$ intersecting five regions $\boldsymbol{A}_i$ $i = 1$ to 5.

We took an arbitrary "cut" from $\boldsymbol{\Omega}$ that we called $\boldsymbol{B}$. The total probability theorem states that the area of $\boldsymbol{B}$ is equal to the sum of the areas that $\boldsymbol{B}$ overlaps with each $\boldsymbol{A}_i$, as shown by the gray areas in Fig. 0.5.

From the definition of the conditional probability and the theorem (0.25), it follows immediately that

$$P(\boldsymbol{A}_i|\boldsymbol{B}) = \frac{P(\boldsymbol{A}_i \cdot \boldsymbol{B})}{P(\boldsymbol{B})} = \frac{P(\boldsymbol{A}_i)P(\boldsymbol{B}|\boldsymbol{A}_i)}{\sum_{j=1}^{n} P(\boldsymbol{A}_j)P(\boldsymbol{B}|\boldsymbol{A}_j)}. \quad (0.26)$$

This is one formulation of Bayes' theorem. It is a very simple theorem, yet it is extremely useful for solving problems that seem intractable.

Suppose that we know the probabilities of all the events $\boldsymbol{A}_i$ (sometimes referred to as *a priori* probabilities) and that we know all the conditional probabilities $P(\boldsymbol{B}|\boldsymbol{A}_i)$. Then Eq. (0.26) allows us to calculate the (*a posteriori*) probabilities $P(\boldsymbol{A}_i|\boldsymbol{B})$. The assignment of the terms "*a priori*" and "*a posteriori*" should not be regarded as ordered in the time. The reason will be clear from the following solution to the urn problem discussed in Sec. 0.5.

We are asked to calculate the probability $P(\text{White}_1|\text{White}_2)$, i.e. given that the outcome on the *second draw* is white, what is the probability that the outcome of the *first draw* is white?

Using Bayes' theorem, we write

$$P(\text{White}_1|\text{White}_2) = \frac{P(\text{White}_1 \cdot \text{White}_2)}{P(\text{White}_2)}$$

$$= \frac{P(\text{White}_2|\text{White}_1)P(\text{White}_1)}{P(\text{White}_2|\text{White}_1)P(\text{White}_1) + P(\text{White}_2|\text{Black}_1)P(\text{Black}_1)}. \quad (0.27)$$

Note that on the right-hand side of (0.27) all the probabilities are supposed to be known. Hence, we can calculate the required conditional probability:

$$P(\text{White}_1|\text{White}_2) = \frac{\frac{1}{3} \times \frac{1}{2}}{\frac{1}{3} \times \frac{1}{2} \times \frac{2}{3} \times \frac{1}{2}} = \frac{1}{3}. \quad (0.28)$$

This is the same as the conditional probability $P(\text{White}_2|\text{White}_1)$.

Exercise 0.1

Generalize the urn problem as in Sec. (0.5) for n white balls and m black balls. Calculate $P(\text{White}_2|\text{White}_1)$ and $P(\text{White}_1|\text{White}_2)$ in the general case. We will further discuss this problem from the information-theoretical point of view in Chap. 3.

0.7 Random Variables, Average, Variance, and Correlation

A *random variable* (rv)[f] is a real-valued function defined on a sample space. If $\boldsymbol{\Omega} = \{w_1, \ldots, w_n\}$, then for any $w_i \in \boldsymbol{\Omega}$ the quantity $X(w_i)$ is a real number.[g] Note that the outcomes of the experiment, i.e. the elements of the sample space, are not necessarily numbers. They

[f]The term "random variable" is actually a function over the domain $\boldsymbol{\Omega}$, assigning to each outcome of the experiment a real value, i.e. $X(w)$ is a real number for each $w \in \boldsymbol{\Omega}$.
[g]We will use the capital letters X and Y for the random variables, but $X(w)$ or $Y(w)$ when referring to the components of these functions. Sometimes we will also use the capital letters X, Y, and Z to denote the experiments.

can be different colors, different objects, different shapes, etc. For instance, the outcomes of tossing a coin are $\{H, T\}$. Another example: suppose we have a die, the six faces of which have different colors, say white, red, blue, yellow, green, and black. In such a case, we cannot plot the function $X(w)$, but we can write the function explicitly as a table. For example, suppose that we have the following values for $X(w)$:

$$\begin{aligned} X(w &= \text{white}) = 2, \\ X(w &= \text{red}) = 2, \\ X(w &= \text{blue}) = 2, \\ X(w &= \text{yellow}) = 1, \\ X(w &= \text{green}) = 0, \\ X(w &= \text{black}) = 1. \end{aligned} \tag{0.29}$$

Note that we always have the equality

$$P(\boldsymbol{\Omega}) = \sum_{w \in \boldsymbol{\Omega}} P(w) = \sum_{i} P\{X(w) = x_i\} = 1. \tag{0.30}$$

The first sum is over all the elements $w \in \boldsymbol{\Omega}$. The second sum is over all values of the rv.

If the outcomes w_i are numbers, and their corresponding probabilities are p_i, then we can define the average outcome of the experiment as usual. For instance, for the ordinary die, we have

$$\sum_{i=1}^{6} w_i p_i = \frac{1}{6} \sum_{i=1}^{6} w_i = 3.5. \tag{0.31}$$

In general, it is meaningless to talk about an average of the outcomes of an experiment when the outcomes are not numbers. For instance, for the six-colored die, there is no average outcome that can be defined for this experiment. However, if an rv is defined on the sample space, then one can always *define* the average of the rv as[h]

$$\overline{X} = \langle X \rangle = E\langle X \rangle = \sum_x x P\{w : X(w) = x\}$$

$$= \sum_w P(w)\, X(w). \tag{0.32}$$

The first sum in Eq. (0.32) is over all possible values of x; the second sum is over all possible outcomes, $w \epsilon \boldsymbol{\Omega}$.

The quantity $\{w : X(w) = x\}$ is an *event*,[i] consisting of all the outcomes w, such that the value of the rv, $X(w)$, is x. In the example (0.29), these events are

$$\{w : X(w) = 0\} = \{\text{green}\},$$
$$\{w : X(w) = 1\} = \{\text{yellow, black}\},$$
$$\{w : X(w) = 2\} = \{\text{white, red, blue}\}. \tag{0.33}$$

[h]We will use the upper case letters X, Y, etc., for the rv, and the lower case letters x, y, etc., for the value of the rv. For an average, we will use either $\overline{X}$ or $\langle X \rangle$. In the mathematical theory of probability, the average is denoted as $E(X)$ and referred to as the "expected value." This is somewhat an unfortunate term for an average. The value of the average is, in general, not an *expected* outcome. For instance, in Eq. (0.31), the average value 3.5 is certainly not an *expected* outcome.

[i]In the mathematical theory of probability, the events are defined as $\{w : X(w) \le x\} \in F$, and the corresponding distribution as $F_X(x) = P\{w : X(w) \le x\}$. In physics, the distribution for an discrete variable is defined as

$$F(x) = P(X = x),$$

and for a continuous rv, the *density* of the distribution is defined such that

$$f(x)\, dx = P(x \le X \le x + dx).$$

Sometimes we will use the shorthand notation for $P\{w : X(w) = x\}$ as $P(X = x)$, or even $P(x)$. If we have several rv's, we will use the notation $P_X(x)$ and $P_Y(y)$ for the distribution of the rv's X and Y, respectively.

The corresponding probabilities are (assuming that the die is "fair")

$$P\{X(w) = 0\} = \frac{1}{6},$$

$$P\{X(w) = 1\} = \frac{2}{6},$$

$$P\{X(w) = 2\} = \frac{3}{6}. \tag{0.34}$$

Hence, the average of the rv's is

$$\overline{X} = 0 \times \frac{1}{6} + 1 \times \frac{2}{6} + 2 \times \frac{3}{6} = \frac{4}{3}. \tag{0.35}$$

The extension of the concept of the rv to two or more rv's is quite straightforward. Let X and Y be the two rv's; they can be defined on the same or on different sample spaces. The joint distribution of X and Y is defined as the probability that X attains the value x, and that Y attains the value y, and is denoted as $P(X = x;\ Y = y)$. The *marginal* distribution of X is defined as

$$P(X = x) = \sum_{y} P(X = x;\ Y = y). \tag{0.36}$$

Similarly, the marginal distribution of Y is defined as

$$P(Y = y) = \sum_{x} P(X = x;\ Y = y). \tag{0.37}$$

The summation is over all y values in Eq. (0.36) and over all x values in Eq. (0.37).

Similarly, for a continuous rv, we replace the distributions by the densities of the distribution, and the marginal distributions are given by

$$f_X(x) = \int f_{X,Y}(x, y)\, dy, \tag{0.38}$$

$$f_Y(y) = \int f_{X,Y}(x, y)\, dx. \tag{0.39}$$

A special average of an rv is the *variance*, which measures how much the distribution is spread out or dispersed. This is defined as

$$\begin{aligned}\mathrm{Var}(X) = \sigma^2 &\equiv \overline{(X-\overline{X})^2} = E(X-E(X))^2 = \overline{X^2} - 2\overline{X}\,\overline{X} + \overline{X}^2 \\ &= \overline{X^2} - \overline{X}^2 \geq 0. \end{aligned} \tag{0.40}$$

For the continuous case, we have

$$\sigma^2 = \int_{-\infty}^{\infty} (x-\overline{X})^2 f_X(x)\, dx \geq 0. \tag{0.41}$$

The positive square root of σ^2 is referred to as the *standard deviation*. Note that from its definition, it follows that σ^2 is always positive. For two random variables X and Y, we define the *covariance* of X and Y as

$$\begin{aligned}\mathrm{Cov}(X,Y) &= E[(X-E(X))\,(Y-E(Y))] = \overline{(X-\overline{X})\,(Y-\overline{Y})} \\ &= \overline{XY} - \overline{X}\,\overline{Y}. \end{aligned} \tag{0.42}$$

The *correlation coefficient* of X and Y is defined as

$$R(X,Y) = \mathrm{Cor}(X,Y) = \frac{\mathrm{Cov}(X,Y)}{\sqrt{\mathrm{Var}(X)\,\mathrm{Var}(Y)}}, \tag{0.43}$$

where the denominator has been introduced to render the range of values of $\mathrm{Cor}(X,Y)$ between -1 and $+1$.

For two independent rv's we have $\mathrm{Cov}(X,Y) = 0$. This follows from the definition of the average

$$\begin{aligned}E(X \cdot Y) = \langle X \cdot Y \rangle &= \sum_{x,y} P(X=x;\ Y=y)\, xy \\ &= \sum_x P(X=x)\, x \sum_y P(Y=y) y = E(X)E(Y) \\ &= \langle X \rangle \langle Y \rangle. \end{aligned} \tag{0.44}$$

Two rv's, for which $E(X \cdot Y) = E(X)\,E(Y)$ are said to be *uncorrelated.*

Note, however, that if $\mathrm{Cov}(X, Y)$ is zero, it does not necessarily follow that X and $Y$are independent. The reason is that independence of X and Y applies for *any* value of x and y, but the uncorrelation applies only to the *average* value of the rv, i.e. $X \cdot Y$. Thus, independence implies that the rv's are uncorrelated, but the converse of this statement is in general not true.

For two uncorrelated random variables X and Y, we have

$$\mathrm{Var}(X + Y) = \mathrm{Var}(X) + \mathrm{Var}(Y). \tag{0.45}$$

It is clear that this relationship holds for two independent rv's. Here, we show that (0.45) is valid under weaker conditions, i.e. that the two rv's are uncorrelated. To show this, we start from the definition of the variance:

$$\begin{aligned}
\mathrm{Var}(X + Y) &= E\{[(X + Y) - E(X + Y)]^2\} \\
&= E[(X - E(X))^2] + E[(Y - E(Y))^2] \\
&\quad + 2E[(X - E(X))\,(Y - E(Y))] \\
&= \mathrm{Var}(X) + \mathrm{Var}(Y).
\end{aligned} \tag{0.46}$$

In the last equality, we have used the condition (0.44), that the two rv's are uncorrelated.

0.8 Some Specific Distributions

We will discuss here, very briefly, a few important distributions that frequently appear in IT.

0.8.1 *The Uniform Distribution*

A discrete rv has a finite number of values, say $(x_1, x_2, \ldots, x_n)$. In such a case, the uniform distribution is defined as

$$P(X = x_i) = \frac{1}{n}, \quad \text{for } i = 1, 2, \ldots, n. \tag{0.47}$$

Here, $(X = x_i)$ is the event $\{w : X(w) = x_i\}$. When X denotes an experiment having n outcomes, we will use the symbol $P(X = x_i)$ to denote the probability that the outcome of the experiment X is x_i. Clearly, Eq. (0.47) defines a probability distribution:

$$\sum_{i=1}^{n} P(X = x_i) = \sum_{i=1}^{n} \frac{1}{n} = 1. \tag{0.48}$$

In the continuous case, the uniform distribution is defined by the density function

$$f(x) = \frac{1}{L}, \quad \text{for } 0 \le x \le L, \tag{0.49}$$

i.e. the probability of finding the outcomes between x and $x + dx$ is constant, independent of x.

Clearly, the normalization of $f(x)$ in Eq. (0.49) is

$$\int_0^L f(x)dx = \frac{1}{L}\int_0^L dx = 1. \tag{0.50}$$

0.8.2 *The Binomial Distribution*

Suppose that we have a series of experiments, each with only two outcomes, say H or T, in tossing a coin, or a particle being in the left or right compartment, or a dipole moment pointing "up" or "down." Let p be the probability of occurrence of one of the outcomes, say H, and $q = 1 - p$ be the probability of the other, say T. If the series of trials are independent, then the probability of any *specific* sequence of outcomes, say

$$\text{H T H H T T H}, \tag{0.51}$$

is simply the product of the probabilities of each outcome, i.e. if $P(H) = p$ and $P(T) = 1 - p = q$, we have, for the specific sequence (0.51) of heads and tails, the probability

$$pqppqqp = p^4 q^3. \tag{0.52}$$

Similarly, if we have seven particles distributed in two compartments, say the right (R) and the left (L), the probability of finding the specific particles (1), (2), (4), and (7) in R and the particles (3), (5), and (6) in L is also p^4q^3.

In most cases, we are not interested in the *specific* configuration, i.e. which particle is in which compartment, but only in the *number* of particle in each compartment.

Clearly, we have many possible *specific* sequences of outcomes for which the number of H's (or of R's) is constant. A specific configuration is written as a sequence, say RLLR, which means the first particle in R, the second particle in L, the third particle in L, and the fourth particle in R. In Table 0.1, we list all the possible *specific* configurations for four particles in two compartments. There is only one

Table 0.1

Specific Configuration	Number of L's	Number of Specific Events in Each Group
RRRR	0	1
LRRR	1	4
RLRR	1	
RRLR	1	
RRRL	1	
LLRR	2	6
LRLR	2	
LRRL	2	
RLLR	2	
RLRL	2	
RRLL	2	
LLLR	3	4
LLRL	3	
LRLL	3	
RLLL	3	
LLLL	4	1

configuration for which all particles are in R, four configurations for which one particle in L, three in R, etc.

Clearly, any two *specific* configurations are disjoint or mutually exclusive events. Therefore, the probability of occurrence of all particles in R is p^4. The probability of occurrence of one particle in L (regardless of which particle is in L) is simply the sum of the four events listed in the table, i.e.

$$P(\text{one particle in } L) = 4p^3q. \tag{0.53}$$

Similarly,

$$P(\text{two particles in } L) = 6p^2q^2 \tag{0.54}$$

and

$$P(\text{three particles in } L) = 4pq^3, \tag{0.55}$$

$$P(\text{four particles in } L) = q^4. \tag{0.56}$$

Thus, in general, for N particles (or N coins), the probability of occurrence of the event "n particles in L" (or n coins showing T) is

$$P_{\mathrm{N}}(n) = \binom{N}{n} p^n q^{N-n}. \tag{0.57}$$

Note that in constructing the expression (0.57), we used the *product rule* for the probabilities of the independent events "R" and "L," and the *sum rule* for the disjoint events which are the "specific sequence" having n in L and $N-n$ in R. The number of specific disjoint events is simply the number of ways of selecting a group of n particles out of N identical particles. The distribution (0.57) is referred to as the *binomial distribution*. The coefficients $\binom{N}{n}$ are the coefficients in the binomial expansion. For any a and b, the binomial expansion is

$$(a+b)^N = \sum_{n=0}^{N} \binom{N}{n} a^n b^{N-n}. \tag{0.58}$$

Sometimes, the distribution $P_N(n)$ is also referred to as the Bernoulli distribution (particularly in the case $p = q = 1/2$.

Figures 0.6 and 0.7 show the binomial distribution $P_N(n)$ defined in Eq. (0.58) for the case $p = q = 1/2$ for different values of N. Note that as N increases, the form of the curve becomes more and more similar to a bell-shaped curve, or the normal distribution (see below).

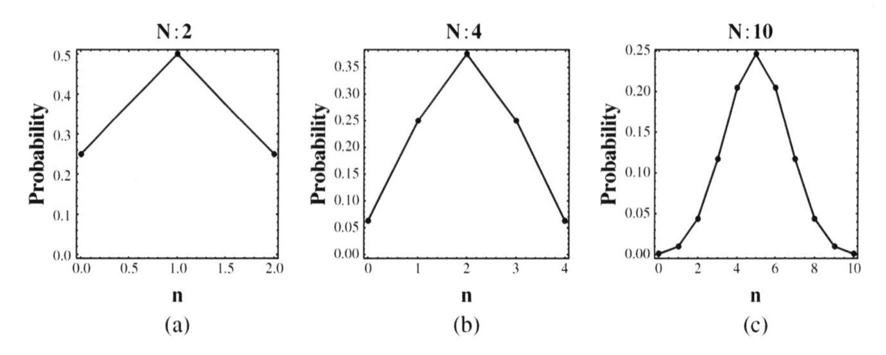

Fig. 0.6. Probability of observing n particles in one compartment and $N-n$ in the other for different numbers N.

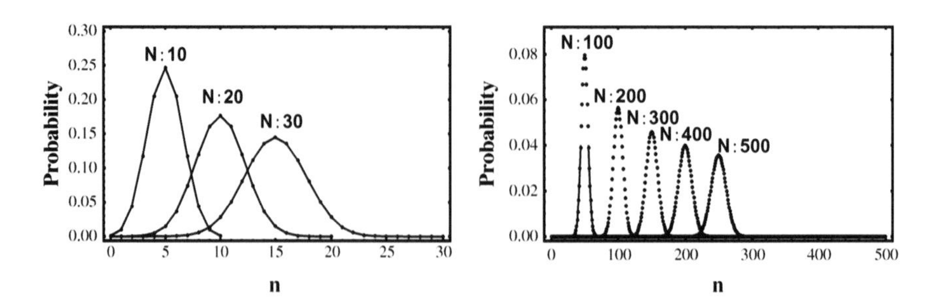

Fig. 0.7. Probability of observing n particles in one compartment and $N-n$ in the other.

Exercise 0.2

Show that the average and the variance of the binomial distribution are

$$E(X) = \sum_{n=0}^{N} nP_N(n) = Np, \tag{0.59}$$

$$\sigma^2 = E(X^2) - [E(X)]^2 = Np\,(1-p). \tag{0.60}$$

Solution

$$E(X) = \sum_{n=0}^{N} nP_N(n) = \sum_n n \binom{N}{n} p^n q^{N-n}, \tag{0.61}$$

where $q = 1 - p$. We will now treat p and q as two independent variables, to write the following identity:

$$\begin{aligned} p\frac{\partial}{\partial p}\left[\sum_n \binom{N}{n} p^n q^{N-n}\right] &= p\sum_n \binom{N}{n} n\,p^{n-1} q^{N-n} \\ &= \sum_n \binom{N}{n} n\,p^n q^{N-n}. \end{aligned} \tag{0.62}$$

Using this identity in Eq. (0.61), we get

$$\begin{aligned} E(X) = \overline{n} = \sum_n nP_N(n) &= p\frac{\partial}{\partial p}\left[\sum_n \binom{N}{n} p^n q^{N-n}\right] \\ &= p\frac{\partial}{\partial p}(p+q)^N = pN(p+q)^{N-1} \\ &= pN. \end{aligned} \tag{0.63}$$

Note that the identity (0.62) is valid for any p and q, whereas in (0.63) we used a particular pair of p and q such that $p + q = 1$.

For the variance, we can use the same trick as above to obtain

$$\sigma^2 = \sum_{n=0}^{N} (n - \overline{n})^2 \; P_N(n) = \overline{(n^2)} - (\overline{n})^2, \tag{0.64}$$

$$\begin{aligned}
\overline{(n^2)} &= \sum_n n^2 P_N(n) = \sum_n n^2 \binom{N}{n} p^n q^{N-n} \\
&= p\frac{\partial}{\partial p}\, p\frac{\partial}{\partial p}\left[\sum_n \binom{N}{n} \; p^n q^{N-n}\right] \\
&= p\frac{\partial}{\partial p}\, p\frac{\partial}{\partial p}\,\left[(p+q)^N\right].
\end{aligned} \tag{0.65}$$

For the particular case $p + q = 1$, we get from Eq. (0.65)

$$\overline{(n^2)} = N(N-1)\,p^2 + Np. \tag{0.66}$$

Hence,

$$\sigma^2 = \overline{(n^2)} - (\overline{n})^2 = N(N-1)\,p^2 + Np - (Np)^2 = Npq. \tag{0.67}$$

0.8.3 *The Normal Distribution*

As we have seen in Figs. 0.6 and 0.7, for very large N the form of the distribution function becomes very similar to the normal, or Gaussian distribution. We will now show that for large N we get the normal distribution as a limiting form of the binomial distribution.

We start with the binomial distribution (0.57) and treat n as a continuous variable. The average and the variance are

$$\overline{n} = \langle n \rangle = \sum_{n=0}^{N} nP_N(n) = pN, \tag{0.68}$$

$$\sigma^2 = \left\langle n^2 \right\rangle - \langle n^2 \rangle = Npq. \tag{0.69}$$

The distribution $P_N(n)$ has a sharp maximum at $\langle n \rangle$; see Figs. 0.6 and 0.7. We expand $\ln P_N(n)$ about the average $\overline{n} = \langle n \rangle$, and take the first few terms,

$$\ln P_N(n) = \ln P_N(\overline{n}) + \frac{\partial \ln P_N(n)}{\partial n}(n-\overline{n}) + \frac{1}{2}\frac{\partial^2 \ln P_N(n)}{\partial n^2}(n-\overline{n})^2 \cdots, \tag{0.70}$$

where the derivatives are evaluated at the point $n = \overline{n}$.

At the maximum, the first derivative is zero. Therefore, we need to consider the expansion (0.70) from the second term. Using the Stirling approximation $\ln N! \approx N \ln N - N$, we can get the second derivative of $P_N(n)$ at $n = \bar{n}$:

$$\frac{\partial^2 \ln P_N(n)}{\partial n^2} = -\frac{1}{Npq}. \tag{0.71}$$

Note that the second derivative is always negative, which means that $P_N(n)$ [as well as for $\ln P_N(n)$] has the maximum at $n = \overline{n}$. The value of the second derivative in Eq. (0.71) is at the point $n = \bar{n}$.

We next rewrite Eq. (0.72) (neglecting all higher order terms in the expansion) as

$$P_N(n) = C \exp\left[-\frac{(n-\overline{n})^2}{2Npq}\right]. \tag{0.72}$$

To normalize this function, we require that

$$\int_{-\infty}^{\infty} P_N(n)dn = 1. \tag{0.73}$$

This integral is well known. The normalization constant is thus

$$C = (2\pi Npq)^{\frac{-1}{2}}. \tag{0.74}$$

Substituting Eq. (0.74) into Eq (0.72), we get

$$P_N(n) = \frac{1}{\sqrt{2\pi Npq}} \exp\left[-\frac{(n-\overline{n})^2}{2Npq}\right]. \tag{0.75}$$

This is the normal, or Gaussian distribution.

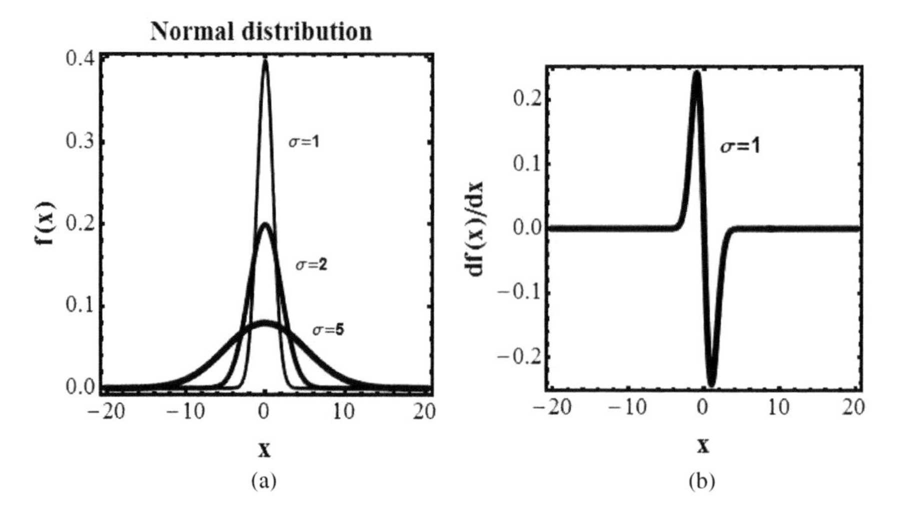

Fig. 0.8. (a) The normal distribution, and (b) its derivative.

Substituting Eq. (0.69),

$$\sigma^2 = Npq, \tag{0.76}$$

we can rewrite this function as

$$P_N(n) \to f(n) = \frac{1}{\sqrt{2\pi\sigma^2}} \exp\left[-\frac{n-\bar{n}}{2\sigma^2}\right]^2. \tag{0.77}$$

Figure 0.8 shows the normal distribution and its derivative. In Sec. 2.2, we will obtain the normal distribution by maximizing the Shannon measure of information.

0.8.4 *The Poisson Distribution*

Another important limit of the binomial distribution is the Poisson distribution. This occurs when $p \to 0$ and $N \to \infty$, but the average, $\lambda = \langle n \rangle = pN$, is constant. In this case, we have

$$\frac{N!}{(N-n)!} = N(N-1)\cdots(N-n+1) \to N^n \tag{0.78}$$

and

$$q^{N-n} = (1-p)^{N-n} \to (1-p)^N = \left(1 - \frac{\lambda}{N}\right)^N \to \exp(-\lambda). \quad (0.79)$$

Hence, we can write Eq. (0.57) in this limit as

$$P_N(n) \to g(n) = \frac{\lambda^n \exp(-\lambda)}{n!}, \quad (0.80)$$

which is the *Poisson distribution.*

It is easy to see that this function is normalized, i.e.

$$\sum_{n=0}^{\infty} g(n) = \sum_{n=0}^{\infty} \frac{\lambda^n \exp(-\lambda)}{n!} = e^{-\lambda} e^{\lambda} = 1. \quad (0.81)$$

The average and the variance of this distribution are

$$\overline{n} = \sum_{n=0}^{\infty} n g(n) = \sum_{n=0}^{\infty} n \frac{\lambda^n \exp(-\lambda)}{n!} = \lambda \exp(-\lambda) \sum_{n=1}^{\infty} \frac{\lambda^{n-1}}{(n-1)!} = \lambda, \quad (0.82)$$

$$\sigma^2 = \overline{(n - \overline{n})^2} = \lambda. \quad (0.83)$$

As an example, suppose that we have a gas in volume V and density $\rho = \frac{N}{V}$. At equilibrium, the density at each point in the system is constant. The probability of finding a specific particle in a small region dV is simply

$$p = \frac{dV}{V}. \quad (0.84)$$

The average number of particles in dV is

$$\lambda = Np = \frac{NdV}{V} = \rho dV. \quad (0.85)$$

If $dV << V$, the probability p is very small. The probability of finding n particles in dV follows the Poisson distribution, i.e.

$$g(n) = \frac{\lambda^n \exp(-\lambda)}{n!}. \quad (0.86)$$

0.8.5 *The Exponential Distribution*

Finally, we mention here one of the most important distributions in physics: the *exponential distribution*. This is defined for any $x \geq 0$ by the density

$$f(x) = \frac{\exp(-\lambda x)}{\int_0^\infty \exp(-\lambda x)\, dx}. \tag{0.87}$$

In Sec. 2.11, we will see how this distribution is obtained by maximizing the Shannon measure of information.

0.9 Bernoulli Trials and the Law of Large Numbers

In this section, we extend the discussion of the binomial distribution discussed in Sec. 0.8.1 in two directions: first, for a very long sequence of experiments; second, for experiments having more than two outcomes.

Let us start with a simple example. We throw a coin n times. We assume that the coin is fair and the results of the experiments are independent. This means that at each instance when the coin is thrown, the probability distribution of the outcomes is the same and independent of the outcomes in the previous experiments. Throwing the coin ten times might give a sequence of results such as

$$\mathrm{H, T, T, H, H, H, T, H, T, T} \tag{0.88}$$

where H stands for "head" and T for "tail". Similarly, the faces of the coin can be labeled with "0" and "1," in which case a possible sequence of ten outcomes could be

$$0, 1, 1, 0, 0, 0, 1, 1, 1, 0. \tag{0.89}$$

In general, we denote the sequence of experiments (or trials, games, or rv's) as $X_1, X_2, \ldots, X_n$.

Fig. 0.9. A one-dimensional system of spins having two orientations; "up" or "down".

The index n is often referred to as indicating the points of times at which the nth experiment is carried out. For instance, we could throw the first coin at time t_1, the second at t_2, and so on. However, n does not necessarily indicate the time at which the nth experiment is carried out. For instance, we have a one-dimensional system, and at each lattice point we have a magnet which can be pointing "up" or "down," in which case the outcome of ten experiments would look like Fig. 0.9.

Thus, the first magnet is at location "1" on the lattice, the second at location "2," and so on. Clearly, in this case, the index n does not signify the time at which the experiment is carried out.

We always assume that the experiments are *independent* and *identically distributed* (i.i.d.). We will use the following notation for the probabilities:

$$P(0) = P(X_i = 0) = p_0,$$
$$P(1) = P(X_i = 1) = p_1. \tag{0.90}$$

In this particular case, $p_0 + p_1 = 1$.

It should be noted that the term "*bit*" was originally defined as "binary – digit." Thus, in the sequence (0.89) we have ten bits. However, in this book we will use the term "bit" as a *unit of information*. In this case, the sequence (0.89) will carry ten bits of information only when the probability of obtaining either "0" or "1" is 1/2 . (See also Sec. 2.2.)

Finally, we consider an experiment which is important in connection with entropy and the second law, which will be discussed in Chap. 5.

Suppose that we have a box with two compartments denoted as R and L; see Fig. 0.10. We throw n labeled marbles into the box (either

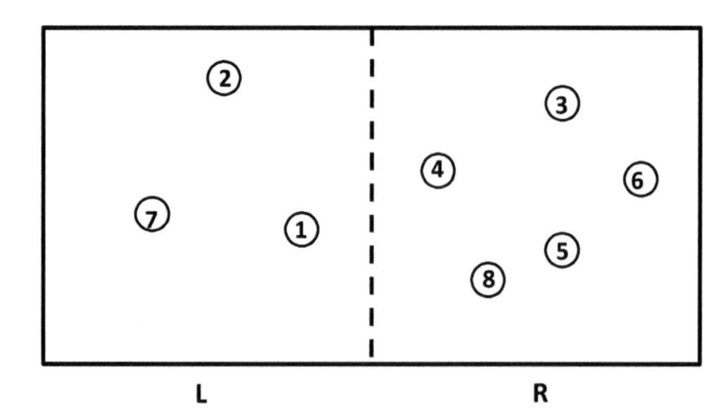

Fig. 0.10. A system with 8 labeled particles in two compartments of equal volumes.

one at a time or all at once), and we assume that the probability of falling into either L or R is 1/2.

We assume that the marbles are independent, i.e. in each experiment a marble can fall into either L or R. With n experiments we might get a sequence of results of the form

$$\mathrm{R, L, R, R, L, R, L, L, L, \ldots} \tag{0.91}$$

As we have done in the case of coins, we can use instead of a sequence of letters a sequence of zeros and ones.

Next, we ask: What is the probability of obtaining a *specific* sequence of zeros and ones, as is (0.89)?

Clearly, since the probabilities of "1" and "0" are equal, and since the n events are independent, the probability of having a *specific* sequence of n_1 ones and n_0 zeros is $(n_1 + n_0 = n)$

$$\left(\frac{1}{2}\right)^{n_1}\left(\frac{1}{2}\right)^{n_0} = \left(\frac{1}{2}\right)^{n_1+n_0} = \left(\frac{1}{2}\right)^{n}. \tag{0.92}$$

We see that in this particular example the probability of obtaining any *specific* sequence is equal to $\left(\frac{1}{2}\right)^n$, independent of n_1 (and $n_0 = n - n_1$). This result might be surprising. Consider the following *specific* sequences with $n = 12$:

$$1, 1, 1, 1, 1, 1, 1, 1, 1, 1, 1, 1;$$
$$1, 0, 1, 0, 1, 0, 1, 0, 1, 0, 1, 0;$$
$$1, 1, 0, 1, 0, 1, 0, 1, 0, 1, 0, 0;$$
$$0, 0, 1, 0, 1, 1, 0, 1, 0, 1, 1, 0. \tag{0.93}$$

One might feel that the probability of obtaining the first two sequences is very small compared with the probability of obtaining either the third or the fourth sequence. This (erroneous) intuitional feeling arises from confusing a *specific* sequence of symbols with a *generic* sequence of symbols.

A *specific* sequence of symbols is a sequence where we *specify* the *location* of each symbol in the sequence (or the location of each specific marble in Fig. 0.10). A *generic* sequence of symbols is a sequence for which we know how many ones and zeros (or R's and L's) there are in the sequence, but we do not care about the specific location of each symbol in the sequence.

What is the probability of finding a *generic* sequence of n_1 ones (hence $n_0 = n - n$ zeros) in n experiments?

To find out the answer, we count how many specific sequences there are with n_1 ones and $n_0 = n - n$ zeros. Clearly, the number of such specific sequences is

$$\binom{n}{n_1} = \frac{n!}{n_0!n_1!}. \tag{0.94}$$

If you have any doubts regarding the number in Eq. (0.94), do the following exercise:

Exercise 0.3

What is the probability of having a *specific* sequence with four symbols, two of which are "1" and two "0"? What is the number of specific sequences with two "1's" and two "0's"?

Solution:

The probability of each specific sequence is $\left(\frac{1}{2}\right)^4$. There are altogether $\binom{4}{2} = 6$ specific sequences with two "1's" and two "0's." These are: 1, 1, 0, 0; 1, 0, 1, 0; 1, 0, 0, 1; 0, 1, 1, 0; 0, 1, 0, 1; 0, 0, 1, 1. Therefore, the probability of the generic sequence is

$$\binom{4}{2} \times \left(\frac{1}{2}\right)^4 = 6 \times \left(\frac{1}{2}\right)^4 = \frac{6}{16} = \frac{3}{8}.$$

The number $\binom{n}{n_1}$ is the number of specific sequences of n_1, "1," and $n_0 = n - n_1$, "0." Since these events are disjoint, the generic sequence is simply the sum over all the corresponding specific sequences. Hence, the probability of finding the generic sequence of n_1, "1," in a sequence of n symbols is

$$P(n_1, n) = \left(\frac{1}{2}\right)^n \binom{n}{n_1}. \tag{0.95}$$

Note that unlike the case of specific sequences, the probabilities of the generic sequences are different. In Figs. 0.6 and 0.7, we have seen some of these probabilities for different values of n. In those figures we used different notations for n and N. We see that for any n this function has a maximum at $n_1 = \frac{n}{2}$. This is the reason why we *feel* that the last two sequences in (0.93) are more "probable" than the first two. Indeed, the first *generic* sequence of $n_1 = 12$ has a probability of $\left(\frac{1}{2}\right)^{12} \approx 0.00024$. On the other hand, all the other sequences in (0.93) have $n_1 = 6$ ones, and hence as *specific* sequences their probabilities are $\left(\frac{1}{2}\right)^{12}$ for each. As generic sequences with $n_1 = 6$, the probability of such a sequence is

$$P(6, 12) = \frac{12!}{6!6!}\left(\frac{1}{2}\right)^{12} \approx 02256. \tag{0.96}$$

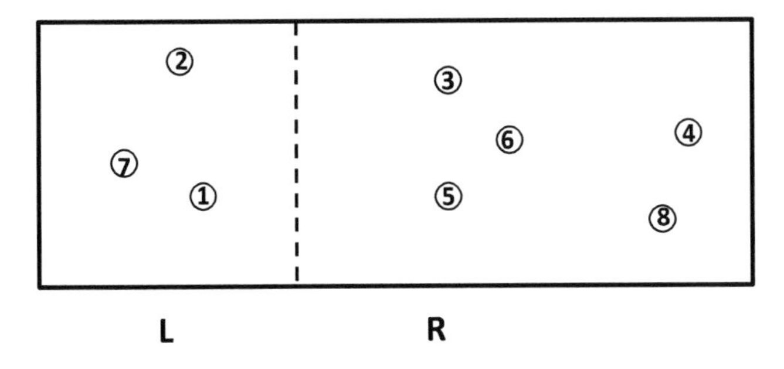

Fig. 0.11. A system with two compartments of unequal volumes.

This is much larger than the probability of the first sequence [note that the probability of the first sequence is $(1/2)^{12}$ whether we view it as specific or generic].

0.9.1 *Generalization for Two Unequal Probabilities*

As in the previous case, we have a sequence of experiments $X_1, X_2, \ldots, X_n$, each with two possible outcomes. One can think of an unfair coin with the probability of H being p_1 and T being p_0. Alternatively, one can think of a board with a total area of one unit divided into two unequal regions with areas $p_L = p_0$ and $p_R = p_1$ (Fig. 0.11), or a box with two compartments having different volumes.

The probability of a *specific* sequence with n_1 ones and n_0 zeros is now

$$P(\text{specific}\, n_1, n_0) = p_1^{n_1} p_0^{n_0} = p_1^{n_1}(1-p_1)^{n_0}. \tag{0.97}$$

Clearly, for $p_1 \neq 1/2$, not all *specific* sequences are equally probable. For instance, the probability of the first specific sequence in (0.93) is p_1^{12}, whereas the probability of the second (or the third or the fourth) sequence is $p_1^6(1-p_1)^6$, which is different from p_1^{12}.

Next, we discuss the *generic* sequence, i.e. a sequence of "1" and "0" such that there are n_0 "0's" (and $n_1 = n - n_0$ "1's"), independently of where the zeros and the ones appear in the sequence. These are sequences having the "structure" as in the second to the fourth sequence in (0.93) — all of these have the same n_0, n_1. Clearly, since the specific sequences with the same n_0, n_1 are disjoint events, we can sum over all the possible specific events to obtain the probability of the *generic* event, i.e.

$$P\left(\text{generic } n_0, n_1\right) = \binom{n_0 + n_1}{n_0} p_0^{n_0} p_1^{n_1}, \tag{0.98}$$

where $p_1 = 1 - p_0$. This is the *binomial distribution*. These terms appear in the binomial expansion

$$(p+q)^n = \sum_{n_0=0}^{n} \binom{n}{n_0} p^{n_0} q^{n-n_0}. \tag{0.99}$$

This expansion is valid for any two numbers q and p.

What is the value of n_0, given n and (p_0, p_1) which maximizes the probability $P\left(\text{generic}, n_0, n_1\right)$? We take the derivative of $\ln P\left(\text{generic}, n_0, n_1\right)$ with respect to n_0, and use the Stirling approximation ($\ln N! \approx N \ln N - N$) to obtain,

$$n_0 = p_0 n. \tag{0.100}$$

This is reminiscent of the relative frequency "definition" of the probability. However, note that in Sec. 0.4 we have *defined* the probability of an event i by

$$p_i = \lim_{n \to \infty} \frac{n_i}{n}. \tag{0.101}$$

In Eq. (0.101), n_i is the number of experiments in which the outcome i has occurred and n is the total number of experiments. Here, p_i is *defined* by this limit. On the other hand, Eq. (0.100) was obtained for n_0, which maximizes the probability $P\left(\text{generic}, n_0, n\right)$, presuming that the probability p_0 (and $p_1 = 1 - p_0$) is given.

0.9.2 *Generalization for Any Number of Outcomes*

Again, we consider a sequence of experiments $X_1, X_2, \ldots, X_n$ which are i.i.d. We denote by p_k the probability of the outcomes A_k:

$$p_k = P(X_i = A_k) \quad \text{for each } i = 1, \ldots, n \quad \text{and each} \quad k = 1, \ldots, c. \tag{0.102}$$

The outcomes A_k are sometimes referred to as being drawn from an alphabet having c letters.

When we perform n experiments, say drawing letters from the English alphabet, we might get a sequence of the form

$$\text{B, D, F, A, K, R, D, E, F.} \tag{0.103}$$

In general, if the outcomes are $A_1, A_2, \ldots, A_c$, we can ask for the probability of obtaining a *specific* sequence characterized by n_1 of the outcome A_1, n_2 of the outcome A_2, and so on. The probability of such a specific sequence is by generalization of (0.97):

$$P(\text{specific sequence with } n_1, n_2, \ldots, n_c) = p_1^{n_1} p_2^{n_2} \ldots p_c^{n_c}, \tag{0.104}$$

where p_i is the probability of the outcome A_i, and n_i is the number of A_i that occurred in the specific sequence. Here, we must have

$$\sum_{i=1}^{c} p_i = 1,$$
$$\sum_{i=1}^{c} n_i = n, \tag{0.105}$$

where n is the total number of outcomes in the sequence.

As in the case of two outcomes, we can ask for the *generic* probability of obtaining a sequence with n_i outcomes A_i, independently of the location of each outcome in the sequence.

This probability is

$$P(\text{generic sequence with } n_1, n_2, \ldots, n_c) = \binom{n}{n_1, n_2, \ldots, n_c} \prod_{i=1}^{c} p_i^{n_i}, \tag{0.106}$$

where the coefficients

$$\binom{n}{n_1, n_2, \ldots, n_c} = \frac{n!}{\prod_{i=1}^{c} n_i!} \tag{0.107}$$

appear in the multinomial expansion

$$(a_1 + a_2 + \cdots a_c)^n = \sum_{\{n_i, n_2, \ldots, n_c\}} \binom{n}{n_1, n_2, \ldots, n_c} \prod_{i=1}^{c} p_i^{n_i}. \tag{0.108}$$

The sum in (0.108) is over all possible sequences $(n_1, n_2, \ldots, n_c)$ such that the sum $\sum_{i=1}^{c} n_i = n$. The vector $(n_1, n_2, \ldots, n_c)$ is referred to as a *partition* of the number n into c integers $n_1, n_2, \ldots, n_c$.

We present here two examples to demonstrate the difference between a specific and a generic sequence.

(a) Suppose that we draw a sequence of letters from an alphabet with three letters: A, B, C. We might obtain sequences of the forms

$$\begin{aligned}
&(n_A = 5; n_B = 3; n_c = 2) \quad \text{B, A, A, B, C, A, C, A, A, B;}\\
&(n_A = 2; n_B = 5; n_c = 3) \quad \text{A, A, C, B, B, C, C, B, B, B;}\\
&(n_A = 2; n_B = 5; n_c = 3) \quad \text{C, A, B, B, C, C, B, B, B, A;}\\
&(n_A = 2; n_B = 5; n_c = 3) \quad \text{B, B, B, A, C, A, C, C, B, B.}
\end{aligned} \tag{0.109}$$

All these are different *specific* sequences. If the probabilities of the letters were equal, i.e.

$$p_A = p_B = p_C = \frac{1}{3},$$

then the probability of each of these specific sequences would be the same, i.e.

$$P(\text{specific}) = \left(\frac{1}{3}\right)^{10}. \qquad (0.110)$$

When the probabilities are different, $p_A \neq p_B \neq p_C$ (such that $p_A + p_B + p_C = 1$), the probability of the first sequence is

$$P\begin{pmatrix}\text{First sequence} \\ \text{in } 0.109\end{pmatrix} = p_A^5 p_B^3 p_C^2. \qquad (0.111)$$

The probabilities of the other three sequences are the same, but different from the probability in (0.111):

$$P\left(\text{the second, third, or the fourth sequence in } 0.109\right) = p_A^2 p_B^5 p_C^3. \qquad (0.112)$$

The generic probability of a sequence having the same partition (n_A, n_B, n_C), but disregarding the specific location of each letter in the sequence, is

$$P\left(\text{generic } (n_A, n_B, n_C)\right) = \frac{n!}{n_A! n_B! n_C!} p_A^{n_A} p_B^{n_B} p_C^{n_C}.$$

(b) The second example is shown in Fig. 0.12. We have three boxes labeled A, B, and C. We distribute n labeled marbles into these boxes such that n_A specific marbles are in A, n_B in B, and n_C in C. The probability of obtaining any *specific* distribution of labeled marbles is

$$P\left(\text{specific } (n_A, n_B, n_C)\right) = p_A^{n_A} p_B^{n_B} p_C^{n_C},$$

and the generic probability is

$$P\left(\text{generic } (n_A, n_B, n_C)\right) = \frac{n!}{n_A! n_B! n_C!} p_A^{n_A} p_B^{n_B} p_C^{n_C}.$$

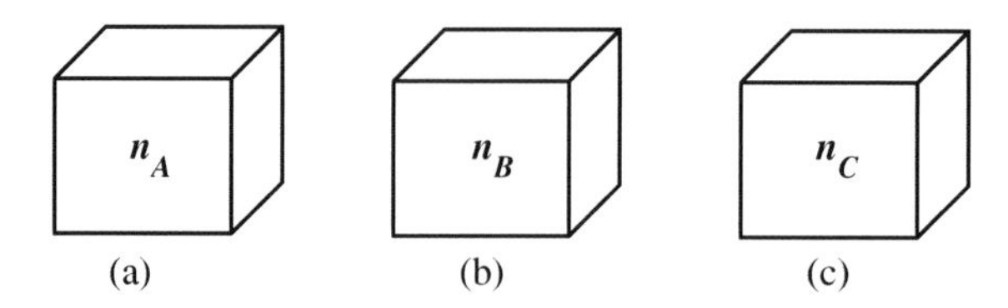

Fig. 0.12. Three boxes, each with a different number of particles.

Note that the specific marble in this case corresponds to the specific *location* of the letter in the previous examples, and the labeled boxes here correspond to the alphabet in the previous example.

0.9.3 *The Law of Large Numbers*

There are several laws of large numbers. We will need only one of them, sometimes referred to as the *weak* law of large numbers.

For a very long sequence of trials, we denote by n_i/n the fraction of the outcomes A_i, having probability p_i. In the limit of very large n we have

$$P\left\{\left|\frac{n_i}{n} - p_i\right| < \varepsilon\right\} \to 1. \qquad (0.113)$$

Intuitively, this law states that as n increases, the probability of the event $\left\{\left|\frac{n_i}{n} - p_i\right| < \varepsilon\right\}$, for any $\varepsilon > 0$, will tend to 1. Qualitatively, this is the main reason for defining the probability p_i as the limit in Eq. (0.9). Note, however, that in (0.9), we *defined* the probability of the event "H" as the limit of a very large number of experiments. In (0.113), we are *given* the probabilities p_i (as postulated by the axiomatic approach to probability discussed in Sec. 0.2), and we are taking the limit of the *distance* between n_i/n and p_i, where p_i is presumed to be given and fulfill the requirements discussed in Sec. 0.2.

In actual application of probabilities, either we *assume* that the probabilities are *given* and that they fulfill the requirement listed in Sec. 0.2, or we do a *finite* number of experiments and *define* the *empirical* probabilities by

$$p_i = \frac{n_i}{n},$$

where n_i is the number of outcomes i that occurred in n experiments.

0.10 The Type of a Sequence and the Typical Sequence

Consider again n experiments, $X_1, X_2, \ldots, X_n$, and the probabilities

$$p(x) = P(X_i = x), \quad \text{for all } i, \tag{0.114}$$

where $x \in \chi$ is a particular result from the alphabet χ.

The *type* of a sequence $\mathbf{x} = x_1, x_2, \ldots, x_n \in \chi^n$ is defined as the sequence of integers $n_1, n_2, \ldots, n_c$, such that n_i is the number of outcomes x_i that occurred in the sequence $\mathbf{x} = (x_1, x_2, \ldots, x_n)$. Note that the *type* of a sequence also defines the *empirical probabilities*:

$$p_{emp} = x_i = \frac{n_i}{n}, \tag{0.115}$$

Thus, in the sequences (0.109), the first has the *type* (5, 3, 2) or, equivalently, the empirical probabilities $\left(\frac{5}{10}, \frac{3}{10}, \frac{2}{10}\right)$. The second, third, and fourth sequences are of the *same type*, i.e. (2, 5, 3) or, equivalently, the empirical probabilities are $\left(\frac{2}{10}, \frac{5}{10}, \frac{3}{10}\right)$.

When we are given the probability distribution $p_1, p_2, \ldots, p_c$, we can write the probability of any sequence of results (of independent experiments, each having the same distribution). Thus, for a sequence of the *type* $(n_1, n_2, \ldots, n_c)$, the probability is

$$P(n_1, n_2, \ldots, n_c) = p_1^{n_1} p_2^{n_2} \cdots p_c^{n_c}. \tag{0.116}$$

We call a sequence of results $x_1, x_2, \ldots, x_n$ *typical* if $n_i \approx np_i$ for all i, or, equivalently, when the *empirical* probabilities n_i/n are nearly equal to the given probabilities p_i.

We denote by $t(n_1, n_2, \ldots, n_c)$ a typical specific sequence (such that $n_i \approx np_i$ for each $i = 1, 2, \ldots, c$). The probability of the typical sequence is obtained from Eq. (0.116), i.e.

$$\begin{aligned} P(t(n_1, n_2, \ldots, n_c)) &= p_1^{n_1} p_2^{n_2} \cdots p_c^{n_c} \\ &\approx p_1^{np_1} p_2^{np_2} \cdots p_c^{np_c} \\ &= 2^{n \sum_{i=1}^{n} p_i \log p_i} = 2^{-nH}, \end{aligned} \qquad (0.117)$$

where $H = \sum_{i=1}^{n} p_i \log_2 p_i$ is the Shannon measure of information, which will be defined in Chap. 1.

Note that t is a specific sequence, i.e. a sequence characterized by $(n_1, n_2, \ldots, n_c)$ in a specific order. The probability of a typical sequence is not necessarily larger than the probability of a non-typical (sometimes referred to as "rare") sequence. For instance, in the example of a three-letter alphabet A, B, C with probabilities $p_A = 0.7, p_B = 0.1$, and $p_C = 0.2$, the probability of the typical sequence (with $n = 10$) is $P(t(n_A = 7; n_B = 1; n_c = 2)) = 0.7 \times 0.1 \times 0.2 \cong 0.00033$, whereas the probability of the nontypical sequence $(n_A = 10; n_B = 0; n_c = 0)$, is $P(n_A = 10; n_B = 0; n_c = 0) = 0.7^{10} = 0.028$, which is much larger than the probability of the typical sequence of length $n = 10$.

There are many important theorems involving typical sequences which we will not discuss here. We note here that if we denote by **T** the union of all typical sequences, then we have

$$P(\mathbf{T}) \approx 1. \qquad (0.118)$$

Denoting by n_T the number of typical sequences, we find that

$$n_T = \frac{P(\mathbf{T})}{P(t)} \approx 2^{nH}, \qquad (0.119)$$

where H is as defined in Eq. (0.117). We will discuss the significance of these results in connection with the second law of thermodynamics in Chap. 5.

0.11 Markov Chains

In the previous section, we discussed a sequence of experiments (or rv's) which are independent. In general, when we have a sequence of experiments, the distribution of outcomes in one experiment depends on the outcomes of all other experiments. In this section, we discuss a particular dependence between the experiments referred to as Markov chains.

Let $(X_1, X_2, \ldots, X_n)$ be a sequence of experiments (or rv's). The joint probability $P(X_1 = x_1, X_2 = x_2, \ldots, X_n = x_n)$ is written in a shorthand notation as $P(x_1, x_2, \ldots, x_n)$, where x_i stands for the event "$X_i = x_i$." The joint probability can always be written in terms of conditional probabilities:

$$\begin{aligned} P(x_1, x_2, \ldots, x_n) &= P\left(x_n \left| x_{n-1}, x_{n-2}, \ldots, x_1\right.\right) P\left(x_{n-1}, \ldots, x_1\right) \\ &= P\left(x_n \left| x_{n-1}, x_{n-2}, \ldots, x_1\right.\right) P\left(x_{n-1} \left| x_{n-2}, \ldots, x_1\right.\right), \\ &\quad P\left(x_{n-2} \left| x_{n-3}, \ldots, x_1\right.\right) \cdots P\left(x_2 \left| x_1\right.\right) P\left(x_1\right). \end{aligned} \tag{0.120}$$

On the right-hand side of Eq. (0.120), we have a product of conditional probabilities. Each term is the conditional probability of obtaining the result x_k given all the previous results. A special case of a sequence of rv's referred to as the *Markov process* (or *discrete Markov process*), when each of the conditional probabilities on the right-hand side of Eq. (0.120) has the form

$$P\left(x_k \left| x_{k-1}, x_{k-2}, \ldots, x_1\right.\right) = P\left(x_k \left| x_{k-1}\right.\right). \tag{0.121}$$

The index n can signify the *time* at which the experiment X_n is carried out, or the point n at which an event has occurred. However, it is common to refer to the Markovian property (0.121) as this: the conditional probability of the event $X_k = x_k$ depends only on the very "recent past" (i.e. $X_{k-1} = x_{k-1}$) and not on the "distant past" (i.e. $X_{k-2} = x_{k-2}, \ldots, X_1 = x_1$).

Note that given the event $X_{k-1} = x_{k-1}$ determines the probability of the event $X_k = x_k$. The conditional probability $P(x_k | x_{k-1})$ is often written as $p_{k-1,k}$, and is referred to as the *transition* probability in *one* step, i.e. from $k - 1$ to k. The transition probability in *two* steps is defined by

$$P(x_k | x_{k-2}) = \sum_{x_{k-1}} P(x_k | x_{k-1}) P(x_{k-1} | x_{k-2}). \tag{0.122}$$

The sum on the right hand side of Eq. (0.122) is over all the possible values of x_{k-1}. Similarly, for the transitional probability in l steps, we have

$$P(x_k | x_{k-l}) = \sum_{x_{k-1} \cdots x_{k-l+1}} P(x_k | x_{k-1}) P(x_{k-1} | x_{k-2}) \ldots$$
$$\times P(x_{k-l+1} | x_{k-l}). \tag{0.123}$$

The summations on the right hand side of Eq. (0.123) is over all the possible values of $x_{k-1}, x_{k-2}, \ldots, x_{k-l+1}$ which are the results of all the "intermediate" experiments between X_{k-l} and X_k.

The reader is urged to pause and think about the meaning of Eqs. (0.122) and (0.123). Equation (0.121) means that if we know that the event $X_{k-1} = x_{k-1}$ occurred, we can calculate the probability of the event $X_k = x_k$. However, if we do not know that the event $X_{k-1} = x_{k-1}$ occurred, but instead we know that the earlier event $X_{k-2} = x_{k-2}$ occurred, then we can calculate the probability of $X_k = x_k$ by summing over all possible intermediate events in Eq. (0.122). A similar meaning applies to Eq. (0.123).

If the transition probabilities depend only on the number of steps l, and not on the index k, the sequence of the experiments is called a *stationary Markov process*, or a *Markov chain*. In such a case, the conditional probability $P_{k-l,k}$ does not depend on k. Specifically, for

$l = 1, P_{k-1,k}$ is independent of k. This property is sometimes referred to as *homogeneity*. This means that the transition probability from the "time" $k - 1$ to the "time" k is independent of the time k. It is simply the transition probability in *one* step. Note again that the index k does not necessarily indicate the time of the experiment. We will see examples of a 1D system where the index k signifies the *location* of a particle in the system.

A property equivalent to the Markovian property is the following:

$$\begin{aligned} &P\left(x_1, x_2, \ldots, x_{k-1}, x_{k+1}, \ldots, x_n \left| x_k\right.\right) \\ &= P\left(x_1, x_2, \ldots, x_{k-1|} \left| x_k\right.\right) P\left(x_{k+1}, \ldots, x_n \left| x_k\right.\right). \quad (0.124) \end{aligned}$$

This equation means that *given* the event $(X_k = x_k)$ makes the events $\left(X_1 = x_1, \ldots, X_{k-1} = x_k\right)$ and $\left(X_{k+1} = x_{k+1}, \ldots, X_n = x_n\right)$ independent. Colloquially, we can say that given the "present" (i.e. $X_k = x_k$) makes the "past" (i.e. $X_1 = x_1, \ldots, X_{k-1} = x_{k-1}$) independent of the "future" (i.e. $X_{k+1} = x_{k+1}, \ldots, X_n = x_n$). The proof of this equivalency is given in App. A. Note again that we are used the terms "present," "past," and "future." These terms are appropriate when the index k signifies the time at which the experiment X_k is (or was) carried out. In general, k does not necessarily signify the time.

We will always assume that all the transition probabilities $P\left(X_k = x_k \left| X_{k-1} = x_{k-1}\right.\right)$ are independent of the index k. We will refer to a discrete *stationary* Markov process as a Markov chain. A Markov chain is completely characterized by an *initial* distribution $\pi^{(0)} = \left(\pi_1^{(0)}, \ldots, \pi_N^{(0)}\right)$ and the transition matrix $\mathbf{P}$, the elements of which are the transition probabilities $p_{i,j}$, a shorthand for $P\left(X_1 = j \left| X_0 = i\right.\right)$. Note here that we changed the notations i and j, instead of x_0 and x_1.

Starting from an initial distribution $\pi^{(0)}$, we can calculate the distribution in the second step as

$$\pi_j^{(1)} = P(X_1 = j) = \sum_{i=1}^{N} P(X_1 = j; X_0 = i)$$
$$= \sum_{i=1}^{N} P\left(X_1 = j \,|X_0 = i\right) P\left(X_0 = i\right)$$
$$= \sum_{i=1}^{N} p_{i,j}\pi_i^{(0)}. \tag{0.125}$$

In the first step in Eq. (0.125), we use the definition of the marginal probability. In the second step, we use the definition of the condition probability, and in the third step we simply use the shorthand notations for the initial probabilities and the transition probabilities. Equation (0.125) can be written in matrix form,

$$\boldsymbol{\pi}^{(1)} = \boldsymbol{\pi}^{(0)}\boldsymbol{P}, \tag{0.126}$$

where on the right hand side we have a multiplication of a row vector by a matrix.

Clearly, one can calculate the distribution in the second step $\pi^{(2)}$, by the same method:

$$\boldsymbol{\pi}^{(2)} = \boldsymbol{\pi}^{(1)}P = \boldsymbol{\pi}^{(0)}\mathbf{P}^2. \tag{0.127}$$

Note that because of the stationary assumption the transition matrix does not depend on the "time," i.e. the index k in Eq. (0.123). Generalizing Eq. (0.127) we have, for the distribution at "time" n,

$$\boldsymbol{\pi}^{(n)} = \boldsymbol{\pi}^{(0)}\boldsymbol{P}^n, \tag{0.128}$$

where $\boldsymbol{P}^n$ is the nth power of P. Note that $\boldsymbol{P} = \boldsymbol{P}^1$ and $\boldsymbol{P}^n$ means matrix multiplication of P, n times, i.e. $\boldsymbol{P}^n = \boldsymbol{P} \times \boldsymbol{P} \times \cdots \boldsymbol{P}$ (n times) is referred to as the transition matrix for n steps. Note also that in some cases the transition probability from i to j in r steps is denoted as $p_{ij}^r = p_{ij}(r)$. What we have seen is that these transition probabilities are given by

$$p_{ij}^r = (\boldsymbol{P}^r)_{ij}. \tag{0.129}$$

On the left hand side of Eq. (0.129), we have the transition probability from i to j in r steps. On the right hand side, we have the ij *element* of the *matrix* P^r. It is easy to show that the transition probability from i to j in $n+m$ steps is given by

$$p_{ij}^{n+m} = \sum_{k=1}^{N} (\boldsymbol{P}^n)_{ik} (\boldsymbol{P}^m)_{kj} = (\boldsymbol{P}^{n+m})_{ij}. \tag{0.130}$$

This equation is known as the Chapman–Kolmogorov equation. The significance of this equation for the case of two steps is simple to visualize; to reach the state j from i in two steps, we need to go through all "paths" leading from i to j. To get the probability of this transition, we need to sum over all intermediate states k.

To conclude this section, we emphasize that Markov chains are only an example — albeit a very useful one — of a sequence of experiments (or rv's). When reading text in any language as a sequence of letters, we should be aware of the fact that the probabilities of finding a specific letter in a specific position, or in a sentence, are neither independent nor obey the Markov properties. In general, the probability of appearance of a letter, say "e" in a word, depends not only on the preceding letter but also on two or more preceding letters in the word.

0.12 The Sum of Independent Random Variables as a Markov Chain

Consider a sequence of rv's $X_1, X_2, \ldots, X_n, \ldots$ which are independent (not necessarily having the same distributions). We define the new sequence of rv's by the partial sums

$$S_1 = X_1,$$
$$S_2 = X_1 + X_2$$
$$\vdots$$
$$S_n = \sum_{i=1}^{n} X_i. \tag{0.131}$$

Clearly, the sequence $\{X_i\}$ defines the sequence $\{S_n\}$. Also, the sequence $\{S_n\}$ defines the sequence $\{X_i\}$.

$$X_1 = S_1,$$
$$X_2 = S_2 - S_1,$$
$$\vdots$$
$$X_i = S_i - S_{i-1}. \tag{0.132}$$

We now show that although the sequence $\{X_i\}$ is independent in the sense that

$$P(X_i, X_j) = P(X_i)\, P(X_j), \tag{0.133}$$

the sequence of $\{S_n\}$ is Markovian.

We have to show that

$$P(S_n = s_n \,|S_{n-1} = s_{n-1}, \dots, S_1 = s_1) = P(S_n = s_n \,|S_{n-1} = s_{n-1}). \tag{0.134}$$

To prove the Markovian property of $\{S_n\}$ in Eq. (0.134), we start with the left hand side and use Eqs. (0.132), i.e. that $X_n = S_n - S_{n-1}$:

$$P(S_n = s_n \,|S_{n-1} = s_{n-1}, \dots, S_1 = s_1)$$
$$= P(S_{n-1} + X_n = s_n \mid S_{n-1} = s_{n-1}, \dots, S_1 = s_1). \tag{0.135}$$

Note that since the condition $S_{n-1} = s_{n-1}$ is on the right hand side of the vertical bar in Eq. (0.135), we can also put it on the left hand side of the bar. In addition, it is clear that if we are given specific values of the $\{X_i\}$, say $X_1 = x_1, X_2 = x_2, \ldots$, these determine uniquely the specific values of the $\{S_i\}$, and vice versa. Therefore, we can rewrite Eq. (0.135) as

$$\begin{aligned} P(X_n &= s_n - s_{n-1} | S_{n-1} = s_{n-1}, \ldots, S_1 = s_1) \\ &= P(X_n = s_n - s_{n-1} | X_{n-1} = S_{n-1} - S_{n-2} \\ &= s_{n-1} - s_{n-2}, \ldots, X_1 = S_1 = s_1). \end{aligned} \quad (0.136)$$

Now, on the right hand side of Eq. (0.136), we have only specific conditions on $\{X_i\}$. But $\{X_i\}$ are all of independent. Therefore, we can delete all of the conditions $X_{n-1}, \ldots, X$ since these do not affect the probability of X_n, and hence we can rewrite Eqs. (0.135) and (0.136) as

$$P\left(S_n = s_n \left| S_{n-1} = s_{n-1}, \ldots, S_1 = s_1\right.\right) = P\left(X_n = s_n - s_{n-1}\right). \quad (0.137)$$

We now start with the right hand side of Eq. (0.134), rewriting it as

$$P\left(S_n = s_n \left| S_{n-1} = s_{n-1}\right.\right) = P\left(S_{n-1} + X_n = s_n \left| S_{n-1} = s_{n-1}\right.\right). \quad (0.138)$$

Since the condition $S_{n-1} = s_{n-1}$ is on the right hand side of the vertical bar, we can also place it on the left hand side to obtain from Eq. (0.138)

$$\begin{aligned} P(S_n &= s_n | S_{n-1} = s_{n-1}) = P\left(X_n = s_n - s_{n-1} \left| S_{n-1} = s_{n-1}\right.\right) \\ &= P\left(X_n = s_n - s_{n-1} \left| X_1 + X_2 + \cdots + X_{n-1} = s_{n-1}\right.\right) \\ &= P\left(X_n = s_n - s_{n-1}\right). \end{aligned} \quad (0.139)$$

The last equality in (0.139) follows from the assumption that all $\{X_i\}$ are independent, and therefore we can delete the conditions on $X_1, X_2, \ldots, X_{n-1}$, which do not affect the probability of X_n.

Comparing Eqs. (0.137) and (0.139), we conclude that the sequence $\{S_n\}$ has the Markovian property (0.134). Thus, we have proven that if we have a sequence of independent experiments or rv's $\{X_i\}$, the partial sums $\{S_n\}$ form a Markov chain.

If in addition we assume that $\{X_i\}$ are identically distributed, then the conditional probability on the right hand side of Eq. (0.134), noting also Eqs. (0.137) and (0.139), is independent of the index n, and therefore the Markov chain $\{S_n\}$ is a *stationary* Markov chain.

1

Introduction, Definition, and Interpretations of Shannon's Measure of Information

In this introductory chapter, we introduce Shannon's measure of information (SMI). We start with Shannon's *motivation* for searching for such a measure. We emphasize, from the outset, that there is an immense difference between the *concept of information* and the *measure of information*. We also note that Shannon erred when he named his measure *"entropy."* This has caused huge confusion in both information theory and thermodynamics. Finally, we survey the various interpretations, as well as a few misinterpretations, of SMI.

1.1 Shannon's Motivation for Constructing a Theory of Information

In 1948, Shannon published a landmark article titled "A Mathematical Theory of Communication." A year later, a slightly expanded work was published as a book by Shannon and Weaver (1949) titled *The Mathematical Theory of Communication*.

Note the minor yet significant difference between the two titles. More importantly, one should note that the article (as well as the book) presents a *"theory of communication"*, not a *"theory of information."* Notwithstanding the difference between "communication" and "information." Shannon's work is now considered to be the cornerstone of what is referred to as "information theory" (IT).

> "Shannon was interested in a theory of *communication of information*, not information itself. This is very clear to anyone who reads through Shannon's original article. In fact, in the introduction to the book, we find the following:
>
> The word information in this theory, is used in a special sense that must not be confused with its ordinary usage. In particular, information must not be confused with meaning.
>
> In fact, two messages, one of which is heavily loaded with meaning and the other of which is pure nonsense, can be exactly equivalent, from the present viewpoint, as regards information. It is this undoubtedly, that Shannon means when he says that "the semantic aspects of communication are irrelevant to the engineering aspects.
>
> The word communication will be used here in a very broad sense to include all of the procedures by which one mind may affect another. This, of course, involves not only written and oral speech, but also music, the pictorial arts, the theatre, the ballet, and in fact all human behavior.
>
> The fundamental problem of communication is that of reproducing at one point either exactly or approximately a message selected at another point. Frequently the messages have meaning ... these semantic aspects of communication are irrelevant to the engineering problem."

It is clear that what is referred to as IT is not a *theory of information!* In fact, there is no theory of information which takes into account the *meaning*, the value, the significance, etc., of the information (see Note 1).

The best way to appreciate what IT is all about is to read Shannon's own words. In Sec. 6 of the article, titled "Choice, Uncertainty and Entropy," we find the following:

"Suppose we have a set of possible events whose probabilities of occurrence are $p_1, p_2, \ldots, p_n$. These probabilities are known but that is all we know concerning which event will occur. Can we find a measure of how much "choice" is involved in the selection of the event or how uncertain we are of the outcome?

If there is such a measure, say, $H(p_1, p_2, \ldots, p_n)$, it is reasonable to require of it the following properties:

(1) H should be continuous in the p_i.
(2) If all the p_i are equal, $p_i = \frac{1}{n}$, then H should be a monotonic increasing function of n. With equally likely events there is more choice, or uncertainty, when there are more possible events.
(3) If a choice be broken down into two successive choices, the original H should be the weighted sum of the individual values of H."

Then Shannon proved the theorem:

"The only H satisfying the three assumptions above is the form: $H = -K \sum p_i \log p_i$.

In this chapter, we will be interested in the *meaning* of SMI as a measure of information. The relevance of H to thermodynamics will be discussed in Chap. 5. Let us quote another paragraph from Shannon's article:

This theorem, and the assumptions required for its proof, are in no way necessary for the present theory. It is given chiefly to lend a certain plausibility to some of our later definitions. The real justification of these definitions, however, will reside in their implications.

Quantities of the form $H = -K \sum p_i \log p_i$ (the constant K merely amounts to a choice of a unit of measure) play a central role in IT as measures of information, choice and uncertainty. The form of H will be recognized as that of entropy as defined in certain formulations of statistical mechanics where p_i is the probability of a system, being in cell i of its phase space. H is then, for example, the H in Boltzmann's famous H theorem. We shall call $H = -K \sum p_i \log p_i$ the entropy of the set of probabilities $p_i, \ldots, p_n$." (See Note 2.)

Before we discuss the various interpretations of H, and before plunging into the applications of this quantity, it is important to pause, read carefully the quotations above, and try to understand what Shannon was aiming at.

First, note that Shannon describes H as a "measure of information, choice, and uncertainty." All these are valid interpretations of the quantity H, as defined above. We will devote more time to these interpretations in the following sections of this chapter.

Second, note that Shannon did not explicitly search for a measure of information. Instead, he formulated his problem in terms of a *probability distribution*, $p_1, \ldots, p_n$. He sought a measure of how much "choice" or "uncertainty" there is in the outcome, and he later referred to the quantity H as a measure of "information, choice, and uncertainty."

Shannon did not seek a measure of the general concept of information, but only a *measure of information contained in, or associated with*, a probability distribution. This is an important point that one should remember whenever using the term "information" either as a measurable quantity or in connection with the second law of thermodynamics (Chap. 5).

Third, Shannon proposed three plausible properties of such a measure, *presuming* that such a measure exists. We will discuss these properties and their plausibility in the following sections of this chapter. Here, the attention of the reader is drawn to the "methodology" of seeking and finding a quantity, the existence of which is not known. Again, we note that the properties specified by Shannon do not apply to the *general concept of information*, but only to a specific type of information.

Finally, note carefully that Shannon was not interested in thermodynamics in general, or *entropy* in particular. However, he noted that "*the form of H will be recognized as that of entropy as defined in certain formulations of statistical mechanics....*" Therefore, he suggested calling H "the entropy of the set of probabilities $p_1, \ldots, p_n$."

Indeed, the *form* of the function H is the same as the *form* of the entropy as used in statistical mechanics. However, this fact *does not* imply that H is entropy. Also, it is not true that H in the Boltzmann H-theorem is entropy. We will further discuss this point in Chap. 5. For the moment, we will study the SMI without any reference to entropy. However, the reader should be aware of the fact that in many applications of the concept of SMI, the concept of entropy has also been involved. This fact has caused great confusion in both IT and thermodynamics.

SMI is a very general concept. It is defined on *any discrete distribution function.* Examples are the outcomes of throwing a die and the frequencies of the appearance of letters in the alphabet of certain languages. There is a vast range of fields in which the quantity H is definable; this has made SMI a very useful tool in many fields of research.

As we will see in Chap. 5, the entropy is defined only on a tiny set of probability distributions. When H is applied to those distributions used in statistical mechanics, it is identical with the statistical-mechanical entropy. Thus, the statistical-mechanical entropy is a particular case of SMI, but the SMI is in general not the entropy. Unfortunately, confusion of the two concepts abounds. The source of this confusion is probably von Neumann's suggestion to Shannon to name the quantity H "entropy." The story is told by Tribus and McIrvive (1971):

> "What's in a name? In the case of Shannon's measure the naming was not accidental. In 1961, one of us (Tribus) asked Shannon what he had thought about when he had finally confirmed his famous measure. Shannon replied: "My greatest concern was what to call

> it. I thought of calling it "information," but the word was overly used, so I decided to call it "uncertainty." When I discussed it with John von Neumann, he had a better idea. Von Neumann told me, 'You should call it entropy, for two reasons. In the first place your uncertainty function has been used in statistical mechanics under that name, so it already has a name. In the second place, and more important, no one knows what entropy really is, so in a debate you will always have the advantage."

In this book, we will refer to the quantity H defined above as the *Shannon's measure of information* (SMI). In App. B, we present a formal proof of the uniqueness of this function. In this and the following chapters, we will survey the properties and the meanings of this quantity as defined above.

Several comments are in order before we discuss the meaning, the properties, and the applications of SMI. The first thing to note is that both in the title of the article and in the introduction, Shannon *was* interested in *communication* theory, not in *information* theory. More specifically, he was interested in transmitting a message from a source to a receiver. The message might or might not have any meaning. The meaning of the *information* carried by the message is irrelevant to the theory he sought to develop.

The second point is that after finding the quantity H which fulfills certain requirements, Shannon says that this quantity "plays a central role in information theory as a measure of information, choice and uncertainty." It should be emphasized that the meaning of the quantity H as a "measure of information" might be misleading. We will discuss shortly in what sense H is a "measure of information."

1.2 Interpretations of SMI

In this section, we discuss three interpretations of SMI. The first is an average of the uncertainty about the outcome of an experiment; the second, a measure of the unlikelihood; and the third, a measure of information. It is ironic that the "informational" interpretation of

SMI is the most difficult to understand, and as a result it is also the one most commonly misused. Other interpretations, such as "choice" on "surprisal" will not be used in this book. We will use the letter H for the quantity defined above, and refer to it simply as SMI. Note that SMI has the form of an average quantity. However, this is a very special average. It is an average of the quantity $-\log p_i$ using the probability distribution $p_1, \ldots, p_n$.

1.2.1 *The Uncertainty Meaning of SMI*

The interpretation of H as an *average uncertainty* is very popular. This interpretation is derived directly from the meaning of the probability distribution.

Suppose that we have an experiment yielding n possible outcomes with probability distribution $p_1, \ldots, p_n$. If, say, $p_i = 1$, then we are *certain* that the outcome i occurred or will occur. For any other value of p_i , we are *less certain* about the occurrence of the event i. *Less certainty* can be translated into "*more uncertainty*." Therefore, the larger the value of $-\log p_i$, the larger the extent of uncertainty about the occurrence of the event i. Multiplying $-\log p_i$ by p_i, and summing over all i, we get an *average uncertainty* about *all* the possible outcomes of the experiment.

We should add here that when $p_i = 0$ we are *certain* that the event i *will not* occur. It would be awkward to say in this case that the *uncertainty* in the occurrence of i is zero. Fortunately, this awkwardness does not affect the value of H. Once we form the product $p_i \log p_i$, we get zero when either $p_i = 1$ or $p_i = 0$.

Yaglom and Yaglom (1983) suggested referring to $-p_i \log p_i$ as the uncertainty about the event i. In this view, the SMI (referred to as "entropy" by Yaglom and Yaglom) is a *sum* over all the uncertainties about the outcomes of the experiment.

This interpretation is invalid for the following reason. As we noted above, it is plausible to interpret $-\log p_i$ as a measure of the extent of uncertainty with respect to the occurrence of the outcome i. Since $-\log p_i$ is a monotonically decreasing function p_i [Fig. 1.1(a)], larger

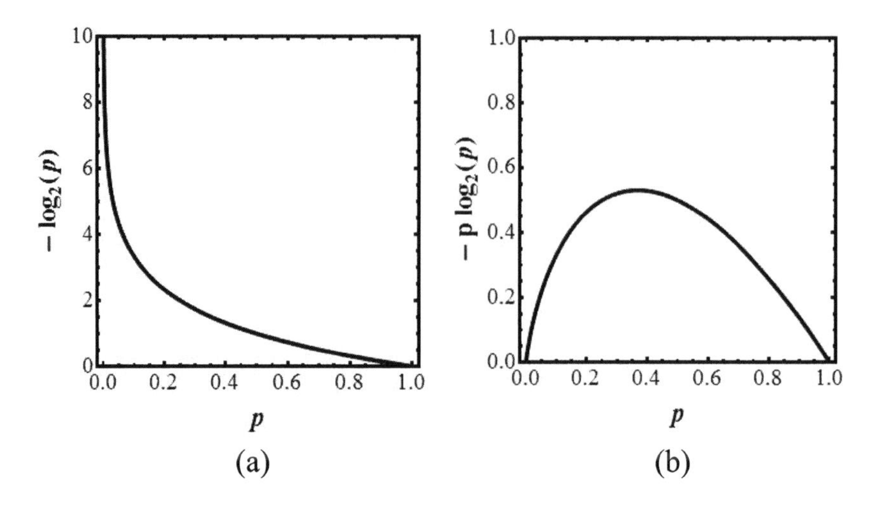

Fig. 1.1. The functions:(a)$-\log(p)$, and (b) $-p\log(p)$.

p_i, or smaller $-\log p_i$, means smaller uncertainty, or larger certainty. In this view, the SMI is an *average* uncertainty over *all* possible outcomes of the experiment.

The quantity $-p_i \log p_i$, on the other hand, is not a monotonic function of p_i [Fig. 1.1(b)]. Therefore, one *cannot* use this quantity to measure the extent of uncertainty with respect to the occurrence of the outcome i.

Finally, we note that whenever we say that SMI is a measure of uncertainty, we mean uncertainty with respect to all the outcomes of an experiment, in the sense discussed above. For instance, when we throw a die, we can talk about many uncertainties; about the color, the mass, the form, etc., of the die. These uncertainties are not the SMI of the experiment of throwing a die. Unfortunately, you can find in many popular science books a description of SMI (as well as entropy) as "uncertainty," without specifying what that uncertainty refers to.

1.2.2 *The Unlikelihood Interpretation*

A slightly different but still useful interpretation of H is in terms of *likelihood* or *expectedness*. These two terms are also derived from the meaning of probability. When p_i is small, the event i is unlikely to occur, or its occurrence is less expected. When p_i approaches 1, the occurrence of i becomes more likely or more expected. Since $\log p_i$ is a monotonically increasing function of p_i, we can say that the larger the value of $\log p_i$, the larger the likelihood or expectedness of the event. Since $0 \leq p_i \leq 1$, we have $-\infty \leq \log p_i \leq 0$. The quantity $-\log p_i$ is thus a measure of the *unlikelihood* or *unexpectedness* of the event i. Therefore, the quantity $H = -\sum p_i \log p_i$ is a measure of the *average unlikelihood*, or *unexpectedness* of the entire set of outcomes of the experiment.

1.2.3 *The Meaning of SMI as a Measure of Information*

As we have seen, both the uncertainty and the unlikelihood interpretation of H are derived from the meaning of the probabilities p_i. The interpretation of H as a measure of information is a little trickier and less straightforward. It is also more interesting since it conveys a different kind of *information* on the Shannon measure of *information*. As we have already emphasized, SMI is not *information*. Also, it is not a measure of any piece of information, but of a very particular kind of information. The confusion of SMI with information is almost the rule, not the exception, for both scientists and nonscientists.

Some authors assign to the quantity $-\log p_i$ the meaning of information (or self-information) associated with the event i.

The idea is that if an event is rare, i.e. p_i is small and hence $-\log p_i$ is large, then one gets "more information" when one knows that the event has occurred. Consider the probabilities of the outcomes of a die as shown in Fig. 1.2(a). We see that outcome "1" is less probable than outcome "2." We may say that we are less uncertain about outcome "2" than about outcome "1." We may also say that outcome "1" is less likely to occur than outcome "2." However, when we are informed that outcome "1" or "2" occurred, we cannot claim that we have received more or less information. When we know that

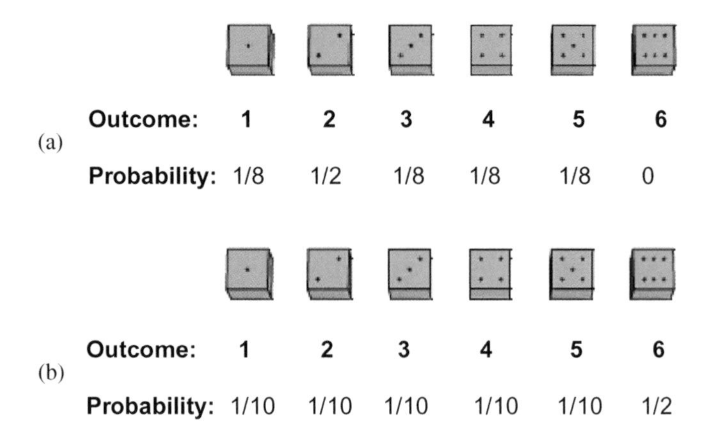

Fig. 1.2. Two possible distributions of an unfair die.

an event i has occurred, we have got the *information* on the occurrence of i. One might be surprised to learn that a rare event has occurred, but the *size* of the *information* one gets when the event i occurs is not dependent on the probability of that event.

Both p_i and $\log p_i$ are measures of the uncertainty about the occurrence of an event. They do not measure *information* about the events. Therefore, we do not recommend referring to $-\log p_i$ as "information" (or self-information) associated with the event i. Hence, H should not be interpreted as *average information* associated with the experiment. Instead, we assign "informational" meaning directly to the quantity H, rather than to the individual events.

It is sometimes said that removing the *uncertainty* is tantamount to obtaining *information*. This is true for the entire experiment, i.e. the entire probability distribution, not individual events.

Suppose that we have an unfair die with probabilities $p_1 = 1/10, p_2 = 1/10, p_3 = 1/10, p_4 = 1/10, p_5 = 1/10$ and $p_6 = 1/2$ [Fig. 1.2(b)]. Clearly, the uncertainty we have regarding the outcome $i = 6$ is less than the uncertainty we have regarding any outcome $i = 6$. When we carry out the experiment and find the result, say $i = 3$, we remove the uncertainty we had about the outcome before carrying out the experiment. However, it would be wrong to argue that the *amount* of information we got is larger or smaller than if another outcome had occurred. Note also that we talk here about the *amount* of information, not the information itself. If the outcome is $i = 3$, the information we got is: the outcome is"3." If the outcome is $i = 6$, the information is: the outcome is "6." These are different information, but we cannot claim that one is larger or smaller than the other.

We emphasize again that the interpretation of H *as average uncertainty or average unlikelihood* is derived from the meaning of each term $-\log p_i$. The interpretation of H as a measure of information is not associated with the meaning of each probability p_i, but with the *entire distribution*, $p_1, \ldots, p_n$.

In this section, we describe in a qualitative way the meaning of H as a *measure of information associated with the entire experiment.* (Note: A measure of information is not *average information.*) We will say more about this interpretation of H in the next chapter.

Consider any experiment or a game having n outcomes with probabilities $p_1, \ldots, p_n$. For concreteness, suppose that we throw a dart at a board, see Fig. 1.3. The board is divided into n regions, of areas $A_1, \ldots, A_n$. We know that the dart hit one of these regions. We assume that the probability of hitting the ith region is $p_i = A_i/A$, where A is the total area of the board.

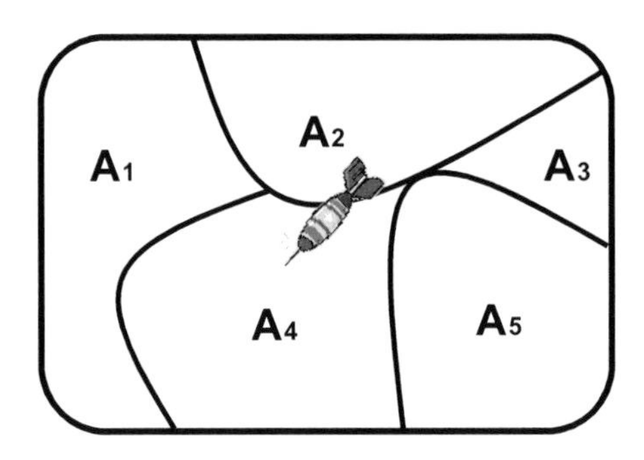

Fig. 1.3. A board divided into five unequal regions.

Now the experiment is carried out, and you have to find out where the dart hit the board. You know that the dart hit the board, and you know the probability distribution $p_1, \ldots, p_n$. Your task is to find out in which region the dart is by asking binary questions, i.e. questions answerable by "yes" or "no".

Clearly, since you do not *know* where the dart is, you *lack information* on the location of the dart. To acquire this information, you ask questions. We are interested in the *amount of information* contained in this experiment. One way of measuring this amount of information is by the *number of questions* you need to ask in order to obtain the required information.

As everyone who has played the 20-question (20Q) game knows, the number of questions you need to ask depends on the *strategy* for asking questions. In the next chapter, we will discuss in more detail what constitutes a strategy for asking questions. For now, we are only interested in a measure of the "amount of information" contained in this experiment. It turns out that the quantity H, which we referred to as Shannon's measure of information, provides us with a measure of this information in terms of the minimum number of questions one needs to ask in order to find the location of the dart, given the *probability distribution* of the various outcomes.

For a general experiment with n possible outcomes, having probabilities $p_1, \ldots, p_n$, the quantity H is a measure of how "difficult" it is to find out which outcome has occurred, given that an experiment was carried out. As we will see in Chap. 2, for experiments having the same total number of outcomes n, but with different probability distributions, the amount of information (measured in terms of the number of questions) is different. In other words, knowing the probability distribution gives us a "hint" or some partial information on the outcomes. This is the reason why we refer to H as a measure of the amount of information *contained in*, or *associated with*, a given probability distribution. We emphasize again that SMI is a measure of information associated with the *entire* distribution, not with the individual probabilities.

1.3 What Shannon Achieved

Recall that Shannon was interested in communication theory, not in information theory. He was interested in the cost, efficiency, time, etc., of transmission of information along communication lines, not in the *meaning* of the information which is transmitted. Thus, the measure of information he obtained was aimed at devising methods for improving the transmission of a sequence of symbols — "improving" in terms of cost, in terms of accuracy in spite of noise, or in terms of efficient codes. In this respect, he achieved his goals. However, in developing the mathematical theory of communication, Shannon achieved much more than what he set out to do. His measure of information, and the other quantities derived from this measure, were found useful in many fields of research which are very remote from, and unrelated to, communication theory.

There are essentially two kinds of applications of SMI. The first is when we know the probability distribution of the outcomes of the experiment. In this case, we can calculate the SMI and use it for interpreting other quantities. Perhaps the most important application of this type is the interpretation of the so-far-mysterious and resistant-to-understanding quantity called *entropy*. We will discuss the power of SMI in demystifying entropy and the second law in Chap. 5.

The second application is the so-called MaxEnt method, which we will refer to as the MaxSMI method. Here, we have an experiment (or a game, or an rv), but we do not know the probability distribution of the outcomes. The question is how to find out the "best" distribution which is consistent with whatever knowledge we have about the experiment.

Finally, the reader should be warned about some unwarranted claims that IT provides solutions to any as-yet unsolved problems.

Contrary to many claims in the literature, IT does not solve any problem. It certainly does not provide any explanation of life, or life-related phenomena such as consciousness, thinking, feeling, or creativity.

1.4 Summary of Chapter 1

In this chapter, we have introduced SMI. For any given distribution $p_1, \ldots, p_n$, the SMI is defined by

$$H = -\sum_{i=1}^{n} p_i \log p_i.$$

The logarithm is to the base 2.

We discussed three interpretations of SMI: as an *average* uncertainty, as an *average* unlikelihood, and as a measure of information. It is important to emphasize that "uncertainty" refers to each outcome, and "unlikelihood" also refers to each outcome, but "measure of information" refers to the whole probability distribution of the outcomes.

The interpretation of H as *average information* does not hold. Instead, the correct interpretation is in terms of the amount of information contained in, or belonging to, the probability distribution of outcomes of an experiment. A simple, informational interpretation of H is in terms of the minimal number of binary questions one needs to ask in order to find a selected event i from the n possible events, given the probability distribution of all the events. This interpretation is discussed in great detail in Chap. 2.

2

Properties of Shannon's Measure of Information

In this chapter, we discuss some of the main properties of SMI. Some of these properties were *assumed* by Shannon when he sought a measure of information, or of uncertainty. In particular, we will see how SMI is related to the number of questions one needs to ask in the familiar 20Q game. We will also introduce the continuous analog of SMI and derive from it three important probability densities.

2.1 Definition of SMI for a Finite Set of Events

As we discussed in Chap. 1, Shannon sought a *measure of information* associated with a probability distribution. The probability distribution is associated with a set of events $A_1, A_2, \ldots, A_n$ which is *complete* and the events are pairwise *mutually exclusive*. This means that one and only one of the events $A_1, \ldots, A_\mathrm{n}$ has occurred, or will occur. The corresponding probabilities are $p_1, \ldots, p_n$ with $p_i \geq 0$ and $\sum_{i=1}^{n} p_i = 1$. Sometimes, we will refer to the distribution associated with an experiment, meaning the probabilities of the outcomes of the experiments.

Thus, for any probability distribution, the SMI is defined by

$$H(p_1, \ldots, p_n) = -\sum_{i=1}^{n} p_i \log p_i. \tag{2.1}$$

The base of the logarithm is usually chosen to be 2. However, in some applications we might use any other base we wish. Note that the quantity H has the *structure* of an average quantity. However, this is a very special kind of average. In general, an average quantity is defined by

$$\langle M \rangle = \sum_{i=1}^{n} p_i M_i. \tag{2.2}$$

Here M_i is some quantity associated with the event A_i, and p_i is the probability of occurrence of this event. The quantity $\langle M \rangle$ is referred to as the *average*, the *mean*, or the *expected* value of the quantity M.

Comparing Eqs. (2.1) and (2.2), we see that H may be interpreted as an average of the quantities $-\log p_i$, each of which itself is related to the probability p_i. In this sense, H may be said to be a *purely probabilistic* quantity.

In Eq. (2.1), we used the notation $H(p_1, \ldots, p_n)$ to denote the functional dependence of H on the variables $p_1, \ldots, p_n$. Sometimes, we use the notation $H(X)$ to denote the quantity H defined for an experiment or an rv X. In more general cases where we have two or more rv's, say X_1, X_2, we denote by $H(X_1, X_2)$ the SMI associated with the two rv's X_1, X_2. In this notation, $H(X_1, X_2)$ does not denote a functional dependence of H on the "variables" X_1, X_2. If p_{ij} is the joint probability of the events $X_1 = x_i$ and $X_2 = x_j$, then $H(X_1, X_2)$ is a shorthand notation for the function of $n \times m$ independent variables p_{ij}:

$$H(p_{11}, p_{12}, \ldots, p_{nm}) = -\sum_{i=1}^{n} \sum_{j=1}^{m} p_{ij} \log p_{ij}. \tag{2.3}$$

Similar generalizations can be applied to any number of rv's. In most of the applications discussed in this book, we define the SMI for a finite number of events (or outcomes). In some applications we use an infinite number of events, or even a continuous case. (See Sec. 2.10.)

2.2 The Case of an Experiment Having Two Outcomes; the Definition of a Unit of Information

The simplest case for which the SMI can be defined is the case of two outcomes, such as the outcomes H and T in the tossing of a coin. This case is important, for several reasons:

(1) It is the simplest case for which we can understand the meaning of H as a measure of uncertainty or as a measure of information.
(2) It is the simplest case with the help of which we can visualize the properties of the function $H(p_1, 1-p_1)$, or simply $H(p)$, where $p = p_1$.
(3) It is the basis for *defining* the *unit of information*, the *bit* (which is different from the bit defined in Sec. 0.9).

From the general definition of H in Eq. (2.1), we can write for the case $n = 2$ the function

$$H(p_1, p_2) = -p_1 \log p_1 - p_2 \log p_2. \tag{2.4}$$

Since $\sum p_i = 1$, H is a function of only one independent variable. We plot this function in Fig. 2.1 For convenience, we write it as

$$H(p) = -p \log p - (1 - p) \log (1 - p). \tag{2.5}$$

This function is defined in the range $0 \leq p \leq 1$. It has a *single* maximum at $p = 1/2$. It is concave downward, and it is zero at both end points: $p = 0$ and $p = 1$.

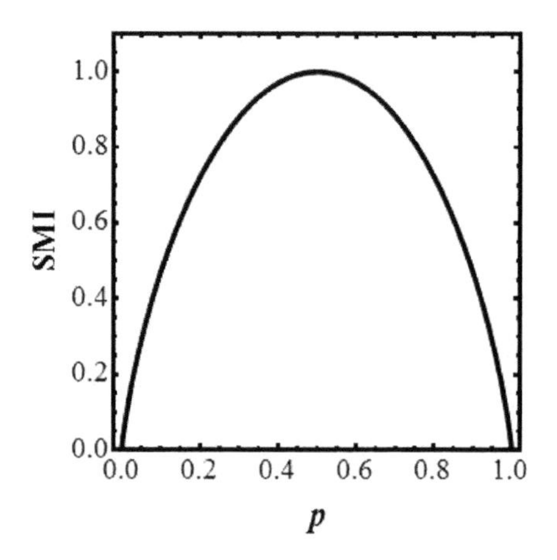

Fig. 2.1. The function $H(p)$ in Eq. (2.5).

The continuity of the function $H(p)$ is obvious. To see that it has a single maximum, we take the two derivatives of $H(p)$ (note that $\log_2 p = \ln p/\ln 2$, where ln is the natural logarithm):

$$(\ln 2)\frac{dH}{dP} = \ln\frac{1-p}{p}, \tag{2.6}$$

$$(\ln 2)\frac{d^2H}{dp^2} = \frac{-1}{p(1-p)}. \tag{2.7}$$

The condition for an extremum is

$$\frac{dH}{dP} = 0 \tag{2.8}$$

or, equivalently,

$$p = (1-p). \tag{2.9}$$

Thus, $H(p)$ has a *single* solution at

$$p = \frac{1}{2}. \tag{2.10}$$

Clearly, the extremum at $p = 1/2$ is a maximum since the second derivative, Eq. (2.7), is always negative in the range $0 < p < 1$.

The value of the function of $H(p)$ at the maximum is

$$H_{\max} = H\left(p = \frac{1}{2}\right) = 1. \tag{2.11}$$

The value of $H(p)$ at each of the extreme points $p = 0$ and $p = 1$ is zero. By using L'Hopital's theorem, it is easy to show that the product $p \log p$ (in any base) tends to zero when p tends to zero:

$$\lim_{p\to 0} (p\ln p) = \lim_{p\to 0} \frac{\ln p}{1/p} = \lim_{p\to 0} \frac{\frac{d}{dp}[\ln p]}{\frac{d}{dp}\left[\frac{1}{p}\right]} = \lim_{p\to 0} \frac{1/p}{-1/p^2} = 0. \tag{2.12}$$

Obviously, the curve in Fig. 2.1 is everywhere concave downward (i.e. the second derivative is negative) in the range $0 \leq p \leq 1$. Before we introduce the definition of the unit of information the bit, it is advisable to consider the following exercises:

Exercise 2.1

A coin is hidden in one of two boxes; see Fig. 2.2. The box in which the coin is hidden was chosen with probabilities as given in Fig. 2.2, cases $a - f$. You are told that the coin is in one of the two boxes, but you *do not know* the probabilities according to which the box was

A	B

Outcome:		A	B	SMI
Probability:	(a)	1	0	0
	(b)	9/10	1/10	0.47
	(c)	2/3	1/3	0.92
	(d)	1/2	1/2	1
	(e)	1/10	9/10	0.92
	(f)	0	1	0

Fig. 2.2. Some values of the function $H(p)$ in Eq. (2.5).

chosen, and you *do not know* where the coin is. Answer the following questions (see Note 1):

(1) What is the uncertainty regarding the location of the coin, i.e. the extent of uncertainty regarding where the coin is, in cases $a - f$?
(2) What is the *information* that you lack regarding where the coin is, in cases $a - f$?
(3) What is the *amount* of information you lack in cases $a - f$?

Exercise 2.2

Next, you are told that, in addition to the fact that there are two boxes and that a coin is hidden in one of them, the boxes were opened and the coin was found in box A. You still *do not know* the probabilities.

Answer the following questions (see Note 2):

(1) How *surprised* were you to know that the coin was found in box A? ("Very surprised," "surprised," "not very surprised," "not surprised".)
(2) How much uncertainty was removed by knowing that the coin was in box A, in cases $a - f$?
(3) What was the *information* you got when you were informed that the coin was found in box A, in cases $a - f$?
(4) How much information did you get ("a lot," "not much," "nothing") when you were informed that the coin was found in box A, in cases $a - f$?

Exercise 2.3

In addition to all the information you got as in Exercise 2.2, you are told that the probabilities are as shown in Fig. 2.2. Answer the following questions, for each of the cases $a - f$ (see Note 3):

(1) How surprised were you to hear that the coin was found in box A?
(2) How much uncertainty was removed after you were informed that the coin was found in box A?

(3) What was the *information* you got when you were informed that the coin was in box A?
(4) How much information did you get when you were informed that the coin was found in box A ("a lot", "not much", "none at all")?
(5) How much uncertainty did you have about the location of the coin before being informed about the location of the coin?
(6) How much information did you lack before you were informed about the location of the coin?

Now that you have a qualitative idea of how to estimate the extent of uncertainty in an experiment, or the amount of information contained in a probability distribution, we define the unit of information, the *bit*, as the *amount of information* you get when you ask a binary question regarding two outcomes having *equal probabilities*, i.e. this is the value of $H(p)$ at $p = \frac{1}{2}, H\left(\frac{1}{2}\right) = -\frac{1}{2}\log\frac{1}{2} - \frac{1}{2}\log\frac{1}{2} = 1$. This is also the maximum information you can get for a binary question; see also Fig. 2.1.

2.3 SMI for a Finite Number of Outcomes

In Chapter 1, we saw that Shannon assumed that a measure of uncertainty or a measure of information must have certain properties. Over the years, various modifications of these properties have been suggested. One convenient list of properties (sometimes referred to as axioms) is the following:

(1) The function $H(p_1, \ldots, p_n)$ is a continuous function of all the variables p_i, where $0 \le p_i \le 1$ and $\sum_{i=1}^{n} p_i = 1$.
(2) When $p_i = p = \frac{1}{n}$ for all $i = 1, 2, \ldots, n$, the function $H(p_1, \ldots, p_n) = f(n)$ is a monotonically increasing function of n.
(3) The function $f(n)$ has the property

$$f(n \times m) = f(n) + f(m). \tag{2.13}$$

(4) The function $H(p_1, \ldots, p_n)$ fulfills the equality

$$H(p_1, \ldots, p_n) = H(p_A, p_B) + p_A H\left(\frac{p_1}{p_A}, \ldots, \frac{p_r}{p_A}\right) + p_B H\left(\frac{p_{r+1}}{p_B}, \ldots, \frac{p_n}{p_B}\right), \quad (2.14)$$

where we have denoted $p_A = \sum_{i=1}^{r} p_i$ and $p_B = \sum_{i=r+1}^{n} p_i$. This property is referred to as the grouping or consistency property.

In App. B, we show that these four properties determine the form of the function H as

$$H = -C \sum p_i \log p_i, \quad (2.15)$$

where C is any positive constant and the logarithm base can be chosen as any number greater than 1. Actually, the condition (2.14) is a special case of a more general grouping condition discussed in App. F of Ben-Naim (2008).

In this section, we show that the function $H(p_1, \ldots, p_n)$ fulfills the four properties.

The continuous property of the function $H(p_1, \ldots, p_n)$ is obvious. Also, for equally probable probabilities $p_i = p = 1/n$, the function

$$f(n) = -\sum_{i=1}^{n} p_i \log p_i = \log n \quad (2.16)$$

which is a monotonically increasing function of n; see Fig. 2.3.

The additivity property

$$f(n \times m) = f(n) + f(m) \quad (2.17)$$

follows from the basic property of the logarithm function, i.e.

$$\log(n \times m) = \log n + \log m. \quad (2.18)$$

The last condition (4) is less intuitive. However, this property is "natural" if we want to interpret H as a measure of information. It is also important in coding theory, but the significance of this property

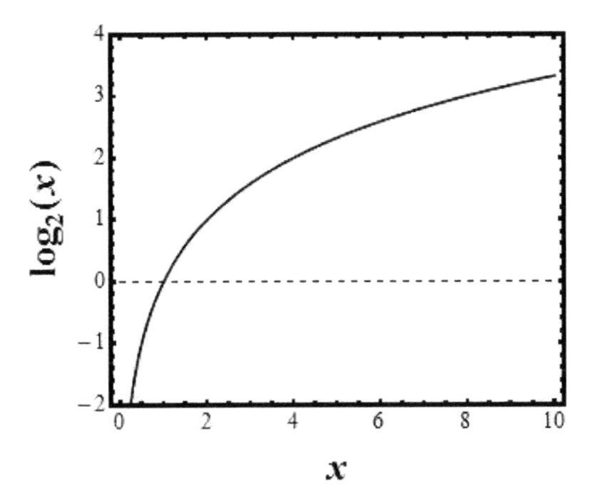

Fig. 2.3. The logarithm as a monotonically increasing function of x.

to coding will not be discussed here. We devote the following section to this property.

2.4 The Special Case of the Grouping Property for the 20Q Game

In this section, we further explore the case of the special grouping into two parts and its relevance to the 20Q game. A more general grouping property is discussed in Appendix F of Ben-Naim (2008).

In fact, the connection between the grouping property and the 20Q game is fundamental to understanding the interpretation of SMI as a measure of information.

First, we show that for any experiment X having n outcomes and a distribution $p_1, \ldots, p_n$, we can write the SMI in the form (2.14). We start with the definition of H for this experiment:

$$H = -\sum_{i=1}^{n} p_i \log p_i. \tag{2.19}$$

We now split all the outcomes into two groups: group A consists of the events $(1, 2, \ldots, r)$, and group B consists of the events

$(r+1,\ldots,n)$. We denote

$$p_A = \sum_{i=1}^{r} p_i, \quad p_B = \sum_{r+1}^{n} p_i, \tag{2.20}$$

and the conditional probabilities:

$$\begin{aligned} P(i \mid A) &= \frac{p_i}{p_A} \quad (\text{for } i = 1,\ldots,r), \\ P(i \mid B) &= \frac{p_i}{p_B} \quad (\text{for } i = r+1,\ldots,n). \end{aligned} \tag{2.21}$$

With these definitions, H in Eq. (2.19) can be written as

$$\begin{aligned} H &= -\sum_{i=1}^{n} p_i \log p_i = -\sum_{i=1}^{r} p_i \log p_i - \sum_{i=r+1}^{n} p_i \log p_i \\ &= -p_A \sum_{i=1}^{r} \frac{p_i}{p_A} \log \frac{p_i}{p_A} - p_B \sum_{i=r+1}^{n} \frac{p_i}{p_B} \log \frac{p_i}{p_B} \\ &\quad - p_A \log p_A - p_B \log p_B \\ &= p_A H\left[P(1 \mid A),\ldots,P(r \mid A)\right] \\ &\quad + p_B H\left[P(r+1 \mid B),\ldots,P(n \mid B)\right] + H\left(p_A, p_B\right). \end{aligned} \tag{2.22}$$

After rearranging the order of terms in Eq. (2.22) and using the definitions in Eqs. (2.20) and (2.21), we obtain Eq. (2.14).

In this section, we examine its relationship to the 20Q game, which is essentially the interpretation of H as a *measure of information*. In Chap. 5, we will also see that this is the basis on which the *informational* interpretation of entropy is established.

Whatever the interpretation of H is — uncertainty, unexpectedness, or a measure of information — Eq. (2.14) tells us that the value of H can be split into two parts: one part is associated with the distribution (p_A, p_B), and the other with the average uncertainty which remains after one finds out which of the events

A and **B** occurred. This is exactly what we do when we play the 20Q game.

2.5 What Is the 20Q Game?

In its most general form, the 20Q game may be described as follows:

We start with n objects (or persons, or events, or whatever it may be). I choose one object, and you have to find out which object I chose by asking binary questions, i.e. questions that are answerable by "yes" or "no". Figure 2.4 shows an example of a 20Q game. We have n objects $A_1, A_2, \ldots, A_n$ and the corresponding probabilities $(p_1, \ldots, p_n)$. Before one asks the first question, one must divide the total of n objects into two groups **A** and **B**, and ask: Is the object in group **A** (or group **B**)? If the answer is "yes", one proceeds to divide the objects in **A** into two groups, and so on. This procedure is shown schematically in Fig. 2.4.

Before we discuss specific games, two comments are in order.

First, in a normal 20Q game we do not explicitly specify all the objects in the set from which the object is chosen. For instance, I can choose a person from the group attending the party, or a person from that city, or a person from the country or from the entire world. Clearly, if I choose a person or an object you have never heard of, you will not be able to find that person or object.

Therefore, in a "fair" game, we must agree on the group of objects or persons from which I chose one. Clearly, this group must be finite,

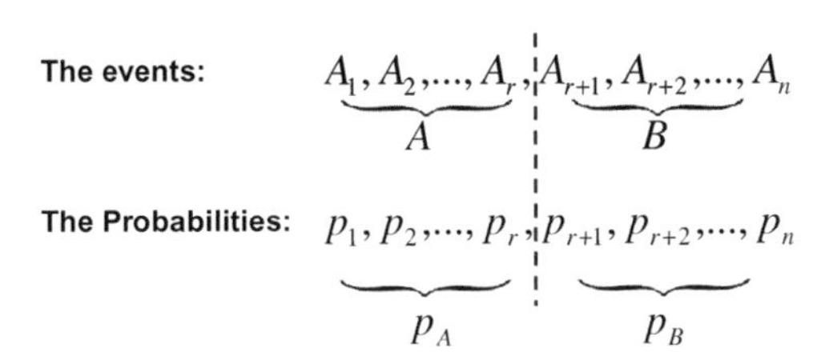

Fig. 2.4. Grouping of n events into two groups; A and B, and their corresponding probabilities.

otherwise there is a chance that you will not be able to find the object in a finite number of questions — and the party can go on forever.

Second, in an actual parlor game, we never specify the probabilities of the events. More precisely, we do not specify with what probability I chose a specific person or object. It is usually assumed implicitly that the objects are chosen with equal probabilities, but in fact that is never the case. For instance, if I had to choose a person, I might be biased to choose a person that I am familiar with, or like, or hate, or someone I know whose identity might be difficult to guess, etc. On the other hand, if you know me, you might use this knowledge to make a better guess about the person I am most likely to choose. Thus, in such a game there are all kinds of psychological elements that can play out into the process of *choosing* the person, as well as of guessing the chosen person. Therefore, in the following, we will make the 20Q game more precise and objective, and hence more susceptible of a mathematical study.

2.6 A Simple 20Q Game

We start with a simple 20Q game. We will analyze two methods or strategies for asking questions, and then calculate the minimum number of questions one needs to ask to guarantee that the required information is obtained.

Consider the game shown in Fig. 2.5. We have four boxes denoted as a, b, c, and d. A coin is hidden in one of them. You know that the probability of finding the coin in a specific box i is 1/4. This is the same as saying that the box in which the coin was placed was selected at random with a *uniform* probability distribution. Your task is to find out where the coin is by asking binary questions.

As anyone who has played the 2Q game knows, there are various strategies for asking questions. It is also intuitively clear that some strategies are "better" than the others, in the sense that they lead to the required information with fewer questions. We will examine these

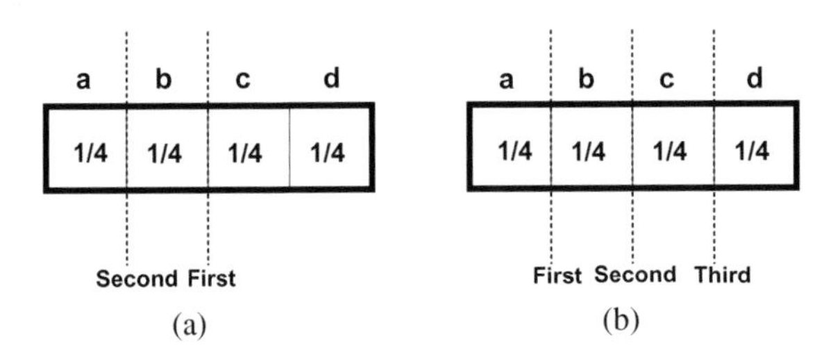

Fig. 2.5. Two strategies of asking questions: (a) The smartest, and (b) the dumbest.

strategies and their differences with a few examples. It is interesting to note that young children learn at an early age the best strategy for asking questions. [For details, see Ben-Naim (2010)].

(i) *The "smartest" strategy*

Everyone who has played the 20Q game knows that it is "inefficient" to ask *specific* questions, such as "Is the person Einstein?" or "Is the person Clinton?" Instead, one has the intuitive feeling that a better strategy for asking questions is to divide the entire pool of objects or persons roughly into two parts, and ask: "Is the person a man [or a woman]?", "Is the person alive [or dead]?", etc. In this manner you *gain* more information at each step of asking questions. We now discuss the most efficient strategy, or the *smartest* strategy, and show in what sense this is the most efficient strategy.

In this strategy, we split the total number of possibilities (outcomes or events) into two groups. Here, the four possibilities are a, b, c, and d, as shown in Fig. 2.5(a). One group consists of the boxes a and b, and the other group consists of the boxes c and d. Based on Eq. (2.14), the first term on the right hand side of this equation corresponds to the question "Is the coin in the group $\boldsymbol{A} = \{\boldsymbol{a}, \boldsymbol{b}\}$?". The answer to this question provides us with one *bit* of information:

$$q_1 = H\left(p_{\mathrm{A}}, p_{\mathrm{B}}\right) = H\left(\frac{1}{2}, \frac{1}{2}\right) = 1 \text{ bit}. \tag{2.23}$$

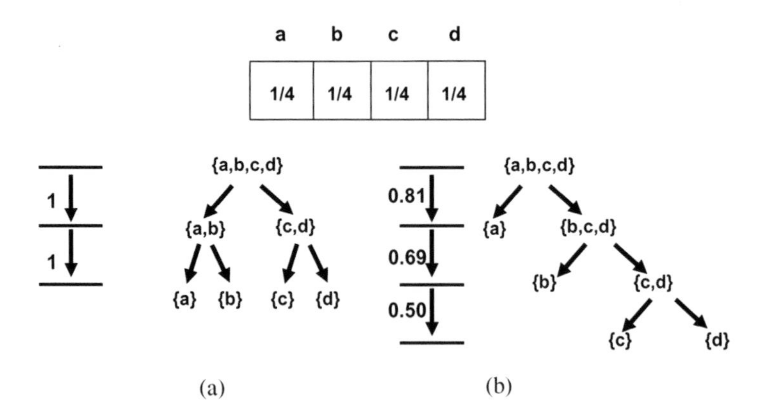

Fig. 2.6. Two strategies of asking questions: (a) The smartest and (b) the dumbest.

Once we get the answer to the first question, we know that the coin is in either $A = \{a, b\}$, or $B = \{c, d\}$. After the first answer is received, the *remaining* SMI consists of the last two terms of Eq. (2.14) which is

$$q_2 = \frac{1}{2}H\left(\frac{1}{2}, \frac{1}{2}\right) + \frac{1}{2}H\left(\frac{1}{2}, \frac{1}{2}\right) = 1 \text{ bit}. \tag{2.24}$$

Thus, in this case, we gain one bit of information from the first question $(q_1 = 1)$, and one bit from the second question $(q_2 = 1)$. [See Fig. 2.6(a).]

The total amount of information we have gained is

$$q_1 + q_2 = 2 \text{ bits}. \tag{2.25}$$

Incidentally, we note that the SMI of this game is

$$H = H\left(\frac{1}{4}, \frac{1}{4}, \frac{1}{4}, \frac{1}{4}\right) = -\sum \frac{1}{4} \log_2 \frac{1}{4} = 2 \text{ bits}. \tag{2.26}$$

We see that the value of H for this game is the same as the sum of the two parts of information $(q_1 + q_2)$ we gained in the two steps.

Although in this example we know that we gain the required information in *exactly* two steps, it is instructive to calculate the *average* number of steps in more detail.

Let us denote by G_i the event "gaining the required information in i steps," and by N_i the event "not gaining the information in i steps."

Clearly, for this strategy there is zero probability of gaining the required information in one step. We write the probabilities of the events G_1 and N_1 thus:

$$P(G_1) = 0,$$
$$P(N_1) = 1. \tag{2.27}$$

Next, we calculate the probabilities of G_2 and N_2:

$$P(G_2) = P(G_2, N_1) = P(G_2 | N_1)P(N_1) = 1 \times 1 = 1. \tag{2.28}$$

This means that in order to gain the information in two steps, we must *not* get the information in the first step, *and* we must get the information in the second step. Using the definition of the conditional probability, we write this joint probability as

$$P(G_2, N_1) = P(G_2 | N_1)P(N_1). \tag{2.29}$$

We already know that $P(N_1) = 1$, i.e. it is certain that we will *not* get the required information in one step. We also know that the conditional probability of G_2 given N_1 is also 1. Therefore, we obtained the result (2.28).

Similarly,

$$P(N_2, N_1) = P(N_2 | N_1)P(N_1) = 0 \times 1 = 0. \tag{2.30}$$

With the two probabilities $P(G_1)$ and $P(G_2)$, we can calculate the average number of steps, i.e.

$$1 \times P(G_1) + 2 \times P(G_2) = 1 \times 0 + 2 \times 1 = 2 \text{ steps}. \tag{2.31}$$

We see that in this case the average number of steps is equal to H [see Eq. (2.26)].

(ii) ***The "dumbest" strategy***

As with the case of the "smartest" strategy, everyone who has played the 20Q game knows that it is not an "efficient" way of asking specific questions, such as "Is the coin in box *a*?" or "Is the coin in box *b*?".

In this section, we discuss the most extreme inefficient strategy, which we refer to as the "dumbest" strategy. Figure 2.5(b) shows this strategy. Here, we also divide the total number of events into two groups: one group consists of a single event, say $\{a\}$, and the other group consists of all the other events, here $\{b, c, d\}$. [See Fig. 2.6(b).]

It is clear that in using this strategy we gain *less* information at each stage of asking questions. Let us see how much information we gain at each step.

In the first step, we ask: "Is the coin in box *a*?". The amount of information we gain after getting the first answer is

$$q_1 = H\left(p_A, p_B\right) = H\left(\frac{1}{4}, \frac{3}{4}\right) \approx 0.8113 \text{ bits.} \tag{2.32}$$

Compare this result with q_1 in the "smartest" strategy, Eq. (2.23). Note that unlike the "smartest" strategy, where you cannot win after the first question, here there is a probability of 1/4 that you get a "yes" answer, and win the game. However, there is a larger probability (3/4) that you get a "no" answer, and you will have to continue asking questions.

In the second step, the average information you get is

$$q_2 = \frac{1}{4}H(1) + \frac{3}{4}H\left(\frac{1}{3}, \frac{2}{3}\right) \approx 0.689 \text{ bits.} \tag{2.33}$$

This is an *average* in the sense that the last two terms of Eq. (2.14) convey average information. Note, however, that in this case we use Eq. (2.14) twice. More specifically, in the first step we have

$$\begin{aligned} H\left(\frac{1}{4}, \frac{1}{4}, \frac{1}{4}, \frac{1}{4}\right) &= H\left(\frac{1}{4}, \frac{3}{4}\right) + \left[\frac{1}{4}H(1) + \frac{3}{4}H\left(\frac{1}{3}, \frac{1}{3}, \frac{1}{3}\right)\right] \\ &= 0.8113 + 1.1887 = 2 \text{ bits.} \end{aligned} \tag{2.34}$$

The first term is the information we gain from the first answer. The second term is the average amount of information left after we receive the first answer. This term is further split into two terms:

$$\frac{3}{4}H\left(\frac{1}{3},\frac{1}{3},\frac{1}{3}\right)=\frac{3}{4}\left[H\left(\frac{1}{3},\frac{2}{3}\right)+\frac{1}{3}H(1)+\frac{2}{3}\left(\frac{1}{2},\frac{1}{2}\right)\right]$$
$$=0.6887+0.5\text{ bits.} \tag{2.35}$$

The first term on the right hand side of Eq. (2.35) is the amount of information we get from the second answer. The second term is the amount of information left after the second answer. In this case, the amount of information we get in the third answer is

$$q_3=\frac{3}{4}\left[\frac{1}{3}H(1)+\frac{2}{3}H\left(\frac{1}{2},\frac{1}{2}\right)\right]=0.5\text{ bits.} \tag{2.36}$$

The sequential reduction in the SMI is shown in Fig. 2.6(b).

The notation $H(1)$ is shorthand for the second term on the right hand side of Eq. (2.14), *where* the group **A** contains only one term, say a. In this case, we have

$$p_A H\left(\frac{p_a}{p_A}\right)=p_A H\left(\frac{p_A}{p_A}\right)=p_A H(1)=0. \tag{2.37}$$

Note again that the sum of the three terms in Eqs. (2.32), (2.33), and (2.36) is

$$q_1+q_2+q_3=2, \tag{2.38}$$

which is the value of H for the entire game.

Next, we turn to calculating the average number of questions we need to ask in this strategy.

The average number of questions in the "dumbest" strategy is obtained from the following probabilities. If we ask "Is the coin in box a?" and obtain the answer "yes", the game is ended. This happens with probability 1/4, and the "no" answer is obtained with probability 3/4. Thus,

$$P(G_1) = \frac{1}{4},$$
$$P(N_1) = \frac{3}{4}. \tag{2.39}$$

Compare these results with the ones in Eqs. (2.27).

To gain the information in the second step, we need to get a "no" answer in the first step, and "yes" in the second step. Hence

$$P(G_2) = P(G_2, N_1) = P(G_2 | N_1)P(N_1) = \frac{1}{3} \times \frac{3}{4} = \frac{1}{4},$$
$$P(N_2, N_1) = P(N_2 | N_1)P(N_1) = \frac{2}{3} \times \frac{3}{4} = \frac{2}{4};$$

and, in the third step,

$$\begin{aligned} P(G_3) &= P(G_3, N_1, N_2) = P(G_3 | N_1, N_2)P(N_1, N_2) \\ &= 1 \times \frac{2}{4} = \frac{2}{4}. \end{aligned} \tag{2.40}$$

The average number of steps in this strategy is

$$\frac{1}{4} \times 1 + \frac{1}{4} \times 2 + \frac{2}{4} \times 3 = \frac{1+2+6}{4} = \frac{9}{4} = 2\frac{1}{4}. \tag{2.41}$$

Note that this is slightly larger than the number of steps in the first, "smartest" method, Eq. (2.31).

From this example we learn that the average number of steps is smaller in the "smartest" than in the "dumbest" strategy. On the other hand, the total SMI of the game is the same, independently of the strategy for asking questions.

Exercise 2.4

Repeat the calculations as in the example of Fig. 2.5 but with eight boxes of equal probability; see Fig. 2.7. Calculate how much

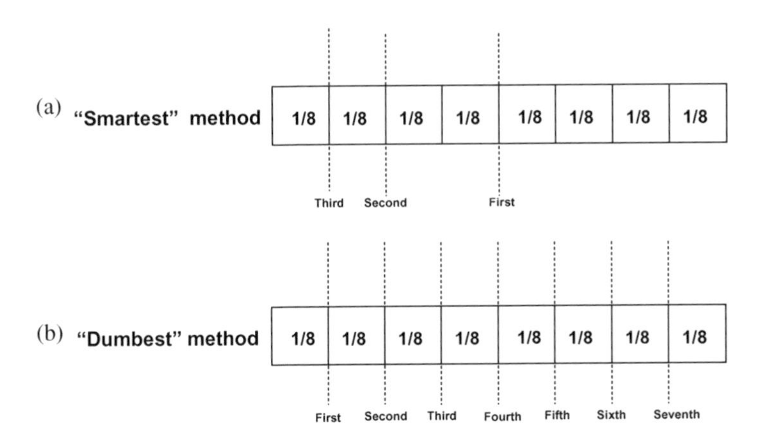

Fig. 2.7. Two strategies of asking questions: (a) The smartest, and (b) the dumbest.

information you get in each step by using the two strategies, and the average number of questions. (For the solution, see Note 4.)

2.7 The General Uniform 20Q Game

Suppose that we have N boxes, in one of which a coin is hidden. In order to find the coin, we ask binary questions. For simplicity, we take N to be of the form $N = 2^n$, where n is a positive integer, and also assume that the probabilities are equal (i.e. uniform distribution). Clearly, there are many ways (or strategies) of asking questions. We will discuss the two extreme cases corresponding to the "smartest" and the "dumbest" strategy.

The total SMI in this case is

$$H\left(\frac{1}{n}, \dots, \frac{1}{n}\right) = \log_2 2^n = n. \tag{2.42}$$

We will study separately the "smartest" and the "dumbest" strategy, as we did in Sec. 2.6.

(i) *The "smartest" strategy*

Using the smartest strategy, the probabilities of gaining the information in the various stages and terminating the game are

$$P(G_1) = 0, \quad P(N_1) = 1,$$

$$P(G_2) = 0, \quad P(N_2) = 1.$$

For $i < n$

$$P(G_i) = 0, \quad P(N_i) = 1.$$

For $i = n$

$$P(G_n) = 1. \tag{2.43}$$

$P(G_i)$ means the probability of gaining the required information in the ith step.

Clearly, on the nth question, we are guaranteed to obtain the required information. This is a straightforward generalization of the results of the previous examples. It is also clear that, since at each step we obtain the maximum information of one bit, we will get the total required information from the minimum number of questions. We can also calculate the average number of questions in this case to be

$$\sum_{i=1}^{n} P(G_i)i = \sum_{i=1}^{n-1} 0 \times i + 1 \times n = n. \tag{2.44}$$

If N is an integer, not necessarily of the form 2^n, we cannot make the same divisions as described above. However, one can always find an integer n such that N will be between the two numbers

$$2^n \leq N \leq 2^{n+1}. \tag{2.45}$$

Thus, if N is the number of boxes, we can increase the number of boxes to the closest number of the form 2^{n+1}. By doing that, the number of questions will increase by at most one. Therefore, the general dependence of the SMI on the number of questions will not change, i.e. for large N the number of questions will be

$$\log_2 N \approx n \log_2 2 = n. \tag{2.46}$$

It should be noted that the above estimates of the number of questions are for N boxes of equal probabilities. This gives an upper limit on the number of questions. If the probability distribution is not uniform, the number of questions necessary for obtaining the information will always be smaller than $\log N$; see below.

(ii) *The "dumbest" strategy*

Using the "dumbest" strategy to ask questions, we have the following probabilities (see also Sec. 2.6). The probability of gaining the information in the first step is simply

$$P(G_1) = \frac{1}{N}, \quad P(N_1) = \frac{N-1}{N}.$$

Note again that, unlike in the "smartest" strategy, there is a finite probability of gaining the information in one step. This probability becomes very small as N increases.

To gain the information in the second step, we must get a "no" answer for the first question, and "yes" for the second, and hence

$$\begin{aligned} P(G_2) &= P(G_2, N_1) = P(G_2 \mid N_1)P(N_1) \\ &= \frac{1}{N-1}\frac{N-1}{N} = \frac{1}{N}, \end{aligned}$$

$$P(N_2, N_1) = P(N_2 \mid N_1)\, P(N_1) = \frac{N-2}{N-1}\frac{N-1}{N} = \frac{N-2}{N},$$

$$P(G_3, N_1, N_2) = P(G_3 \mid N_1, N_2)\, P(N_1, N_2) = \frac{1}{N-2}\frac{N-2}{N} = \frac{1}{N},$$

$$P(N_3, N_1, N_2) = P(N_3 | N_1, N_2) P(N_1, N_2)$$
$$= \frac{N-3}{N-2} \frac{N-2}{N} = \frac{N-3}{N},$$
$$\vdots$$
$$P(G_{N-1}) = P(G_{N-1}, N_1, \ldots, N_{N-2}) = P(G_{N-1} | N_1, \ldots, N_{N-2})$$
$$\times P(N_1, \ldots, N_{N-2}) = 1 \times \frac{2}{N} = \frac{2}{N}. \quad (2.47)$$

To get the information in the $(N-1)$th step, we need to get "no" answers for all the previous $(N-2)$ questions, and a "yes" answer for the $(N-1)$th question.

At the $(N-1)$th answer, we get the required information. The average number of questions in this strategy is

$$\sum_{i=1}^{N-2} \frac{i}{N} + (N-1)\frac{2}{N} \xrightarrow{\text{for large } N} \frac{N}{2}. \quad (2.48)$$

Thus, for large N the average number of questions in this case is *linear* in N, whereas in the smartest strategy it depends on N as $\log_2 N$. This difference in behavior is shown in Fig. 2.8.

Check that this expression gives the same average number of questions for the cases $N = 4$ and $N = 8$.

2.8 The Nonuniform Game

In this section, we discuss a simple nonuniform 20Q game. Figure 2.9 shows a game with four outcomes $\{a, b, c, d\}$ having the probabilities

$$P(a) = \frac{1}{8}, \ P(b) = \frac{1}{8}, \ P(c) = \frac{1}{4}, \ P(d) = \frac{1}{2}. \quad (2.49)$$

You can think of this game as in the case of Fig. 2.6, i.e. we have four boxes, and a coin is hidden in one of the boxes. However, unlike the game in Fig. 2.6, here the choice of the box is made with different probabilities.

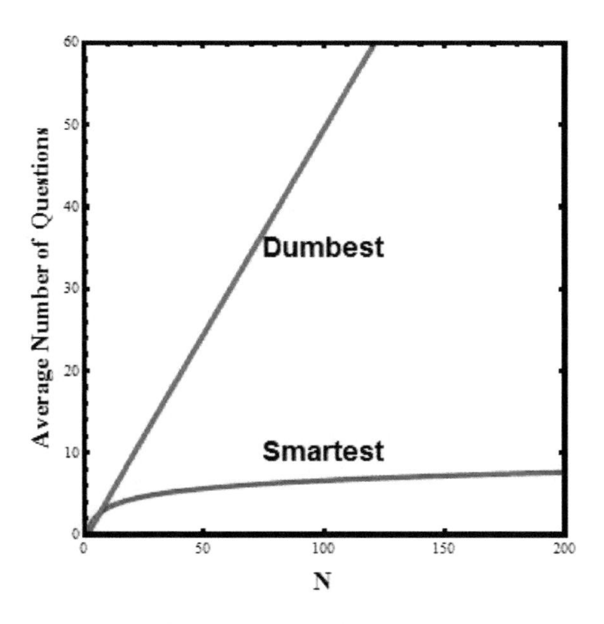

Fig. 2.8. Dependence of the number of questions one needs to ask as a function of N for the two strategies.

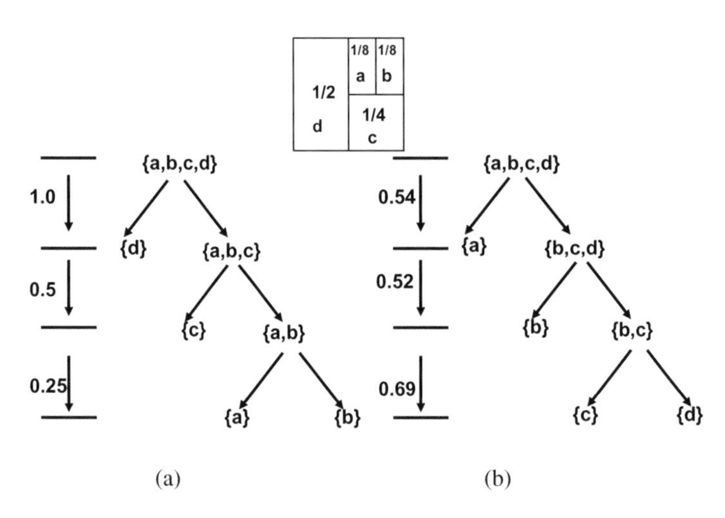

Fig. 2.9. Two strategies of asking questions: (a) The smartest, and (b) the dumbest.

A simpler way of visualizing the probability distribution is to think of a board with a total area of unity. This board is divided into four different regions with corresponding areas 1/8, 1/8, 1/4 and 1/2. While blindfolded, we throw a dart at the board. We are told that the dart hit the board, and we know the areas of the four regions. We assume that all the points of the board are equally probable, and therefore the probability that the dart hits the region i will be proportional to the area of that region. Since we chose the total area of the board to be unity, the relative areas of the region are also equal to the probabilities of finding the dart in the different regions (we neglect the probability that the dart hits exactly the line bordering two adjacent regions).

The value of the SMI for this game is

$$\begin{aligned} H &= -\left(\frac{1}{2}\log_2\frac{1}{2}+\frac{1}{4}\log_2\frac{1}{4}+\frac{2}{8}\log_2\frac{1}{8}\right)=\frac{4+4+6}{8} \\ &= \frac{7}{4}=1\frac{3}{4}=1.75 \text{ bits.} \end{aligned} \tag{2.50}$$

(i) ***The "smartest" method***

The average gains of information at each step are shown in the left diagram of Fig. 2.9(a), and the arrows show the reduction in the SMI at each step:

$$\begin{aligned} q_1 &= H\left(\frac{1}{2},\frac{1}{2}\right)=1 \text{ bit}, \\ q_2 &= \frac{1}{2}H(1)+\frac{1}{2}H\left(\frac{1}{2},\frac{1}{2}\right)=\frac{1}{2}\text{ bit}, \\ q_3 &= \frac{1}{2}\left[\frac{1}{2}H(1)+\frac{1}{2}H\left(\frac{1}{2},\frac{1}{2}\right)\right]=\frac{1}{2}\times\frac{1}{2}=\frac{1}{4}\text{ bit.} \end{aligned} \tag{2.51}$$

Hence, the total amount of information gained in the three steps is

$$q_1+q_2+q_3=1\frac{3}{4}=1.75 \text{ bit}, \tag{2.52}$$

which is the same as the SMI in Eq. (2.50).

The probabilities for this case are

$$P(G_1) = \frac{1}{2},$$

$$P(N_1) = \frac{1}{2},$$

$$P(G_2, N_1) = P(G_2\,|N_1\,)\,P(N_1) = \frac{1}{2} \times \frac{1}{2} = \frac{1}{4}. \qquad (2.53)$$

The average number of steps is

$$\frac{1}{2} \times 1 + \frac{1}{4} \times 2 + \frac{1}{4} \times 3 = \frac{2+2+3}{4} = 1\frac{3}{4} = 1.75, \qquad (2.54)$$

which is the same as the total SMI in Eq. (2.50).

(ii) ***The "dumbest" method***

The average gain of information at each step [see diagram in Fig. 2.9(b)], is

$$q_1 = H\left(\frac{1}{8}, \frac{7}{8}\right) = 0.544\,\text{bit},$$

$$q_2 = \frac{1}{8}\,H\,(1) + \frac{7}{8}\,H\left(\frac{1}{7}, \frac{6}{7}\right) = 0.518\,\text{bit},$$

$$q_3 = \frac{7}{8} \times \frac{6}{7} H\left(\frac{1}{3}, \frac{2}{3}\right) = 0.689\,\text{bit}. \qquad (2.55)$$

The sum of these gains of information is

$$q_1 + q_2 + q_3 = 1.75\,\text{bit}, \qquad (2.56)$$

which is again the same as in Eq. (2.50).

The probabilities for this strategy are

$$P(G_1) = \frac{1}{8},$$

$$P(N_1) = \frac{7}{8},$$

$$P(G_2, N_1) = P(G_2 \,|N_1\,) P(N_1) = \frac{1}{7} \times \frac{7}{8} = \frac{1}{8},$$
$$P(N_2, N_1) = P(N_2 \,|N_1\,) P(N_1) = \frac{6}{7} \times \frac{7}{8} = \frac{6}{8},$$
$$P(G_3, N_1, N_2) = P(G_3 \,|N_1, N_2\,)\, P(N_1,\, N_2) = 1 \times \frac{6}{8} = \frac{6}{8}. \quad (2.57)$$

The average number of steps is

$$\frac{1}{8} \times 1 + \frac{1}{8} \times 2 + \frac{6}{8} \times 3 = \frac{1 + 2 + 18}{8} = 2\frac{5}{8}, \quad (2.58)$$

which is *larger* than in the "smartest" method.

From this example, we can make the following conclusions: first, the number of steps in the smartest method is smaller than in the "dumbest" method. Second, the number of steps, or the number of questions, in the "smartest" method is the same as the value of the SMI for the same game. Third, the SMI as well as the number of questions in the "smartest" method for the nonuniform distribution is smaller than the number of questions for the uniform game with the same total number of events (here $N = 4$).

We will see in the next section that all of these conclusions are valid for any game.

Exercise 2.5

Three boxes with unequal probabilities.

Consider the game described in Fig. 2.10. The probabilities of the boxes are $1/4, 1/4, 1/2$. Calculate all the relevant quantities as we have done for the game in Fig. 2.9. (For the solution, see Note 5.)

2.9 The Maximum of the SMI Over All Possible Discrete Distributions

We now generalize what we have found in the previous section. We are given a game (or an rv, or an experiment) having n outcomes

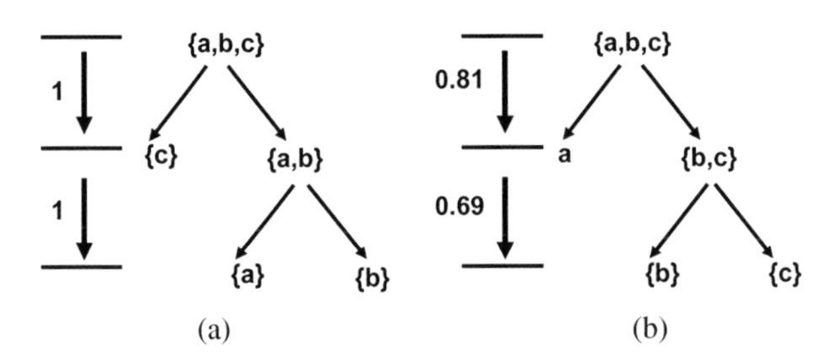

Fig. 2.10. Two strategies of asking questions: (a) The smartest, and (b) the dumbest.

with a distribution $p_1, \ldots, p_n$. The function $H(p_1, \ldots, p_n)$ has a *single maximum* over all possible distributions with the same n.

The mathematical problem is to find the maximum of the function

$$H(p_1, \ldots, p_n) = -\sum_{i=1}^{n} p_i \log p_i, \tag{2.59}$$

subject to the constraint (or the closure condition)

$$\sum_{i=1}^{n} p_i = 1. \tag{2.60}$$

In this and in the next section, we use the natural logarithm for convenience. This choice simplifies the mathematics but does not affect the final result.

We define the auxiliary function

$$F = H\,(p_1, \ldots, p_n) + \lambda \left(\sum_{j=1}^{n} p_j - 1 \right), \tag{2.61}$$

where λ is a constant which will be determined later. The condition for an extremum of F is that all the partial derivatives of F with respect to each of the p_i are zeros, i.e.

$$\left(\frac{\partial F}{\partial p_i}\right)_{p_i'} = -\log p_i - 1 + \lambda = 0. \tag{2.62}$$

The symbol p_i' stands for the vector $(p_1, p_2, \ldots, p_{i-1}, p_{i+1}, \ldots, p_n)$, i.e. all the components except p_i. From Eq. (2.62), we get the distribution which maximizes H:

$$p_i^* = \exp(\lambda - 1). \tag{2.63}$$

Substituting Eq. (2.63) into Eq. (2.60), we obtain

$$1 = \sum_{i=1}^{n} p_i^* = \exp(\lambda - 1) \sum_{i=1}^{n} 1 = n \exp(\lambda - 1). \tag{2.64}$$

Hence, we have

$$p_i^* = \frac{1}{n}. \tag{2.65}$$

This is an important result. It says that the maximum value of H, subject to only the condition (2.60), is obtained when the distribution is *uniform*. This is a generalization of the result we have seen in Sec. 2.2. It is easy to see that at the uniform distribution H has a *maximum*. This follows from the fact that the second derivative of H is always negative:

$$\frac{d^2 H}{dP_i^2} = \frac{-1}{P_i} < 0 \quad \text{(for all } i\text{)}. \tag{2.66}$$

The value of H at the maximum is

$$H_{\max} = -\sum_{i=1}^{n} p_i^* \log p_i^* = -\sum_{i=1}^{n} \frac{1}{n} \log \frac{1}{n} = \log n. \tag{2.67}$$

Clearly, when there are n equally likely events, the value of H is larger, the larger the number of possible outcomes.

Exercise 2.6

Calculate the minimum number of questions you need to ask, in order to find the box in which the coin is located for $n = 8, 16, 32$. Why is it that when we *double* the number of boxes we add only one more question? (See Note 6.)

2.10 The Case of Infinite Number of Outcomes

The definition of SMI for the case of a discrete infinite number of possibilities is straightforward. First, we recall that for a finite and uniform distribution, we had

$$H = \log n, \tag{2.68}$$

where n is the number of possibilities. Taking the limit, $n \to \infty$, we get

$$H = \lim_{n \to \infty} \log n = \infty. \tag{2.69}$$

This means that the amount of information contained in the game tends to infinity. Note, however, that the probabilities $1/n$ tend to zero. The interpretation of this result is simple. If we have an infinite number of equally probable possibilities, then we will need on average an infinite number of questions to ask.

For nonuniform distribution, the quantity H might or might not exist, depending on whether the quantity

$$H = -\sum_{i=1}^{\infty} p_i \log p_i \tag{2.70}$$

converges or diverges.

The case of a continuous distribution is problematic. If we start from the discrete case and proceed to the continuous limit, we get into some difficulties. We will discuss this problem in App. C. Here, we will follow Shannon's treatment for a continuous distribution for which a density function $f(x)$ exists. In analogy with the definition of the H function for discrete probability distribution, we define the quantity H for a continuous distribution. Let $f(x)$ be the density distribution, i.e. $f(x)dx$ is the probability of finding the rv having values between x and $x + dx$.

We defined the H function as

$$H = -\int_{-\infty}^{\infty} f(x) \log f(x) dx. \tag{2.71}$$

Note carefully that in this definition $f(x)dx$ is a pure number. In general, $f(x)$ itself is not a pure number [for example, if dx has units of length, then $f(x)$ has units of 1/length]. Therefore, one must be careful in using this definition of H. (See also Appendix C.)

As we noted in the previous section, for mathematical convenience we use in this section the natural logarithm in the definition of H.

2.11 Three Extremum Theorems on SMI

In this section, we discuss three important theorems proven by Shannon (1948). They are important for three reasons. First, they show how three fundamental distributions in probability theory and statistics arise. Second, they shed new light on the meaning of the equilibrium state in thermodynamics. Third, they are essential to the understanding of both entropy and the second law of thermodynamics. (See Chapter 5.)

2.11.1 *The Uniform Distribution of Locations*

Consider a particle that is confined to a 1D "box" of length L. We seek the maximum of H defined in Eq. (2.71), but with limits $(0, L)$, subject to the conditions that

$$\int_{0}^{L} f(x) dx = 1. \tag{2.72}$$

We use the Lagrange method of the undetermined multiplier (or the calculus of variation). We define the auxiliary functional

$$A\left[f(x)\right] = H\left[f(x)\right] + \lambda \int_0^L f(x)dx. \tag{2.73}$$

Taking the functional derivative with respect to the component $f(x')$, we obtain

$$\frac{\delta A}{\delta f\,(x')} = -1 - \log f(x') + \lambda. \tag{2.74}$$

For details, see Appendix D. From Eq. (2.74), we find that the probability density $f^*(x)$ which maximizes H, subject to the condition (2.72), must satisfy the equality

$$-1 - \log f^*(x) + \lambda = 0. \tag{2.75}$$

Using the result of Eq. (2.75) in Eq. (2.72), we obtain

$$1 = \int_0^L f^*(x)dx = e^{\lambda-1} \int_0^L dx = e^{\lambda-1}L \tag{2.76}$$

or, equivalently,

$$f^*(x) = \frac{1}{L}. \tag{2.77}$$

This is the density which maximizes H subject to the condition (2.72). We will refer to this distribution as the *equilibrium distribution* (see Chapter 5), and use the notation $f_{eq}\,(x)$ instead of $f^*\,(x)$.

Thus, the equilibrium density distribution is *uniform* over the entire length, L. The probability of finding the particle at any interval, say between x and $x + dx$, is

$$f_{eq}\,(x)\,dx = \frac{dx}{L}, \tag{2.78}$$

which is independent of x. This result is of course in accordance with our expectations. Since no point in the box is preferred, the probability of being found in an interval dx is simply proportional to the length of that interval. A more general result is when the density function

$f(x)$ is defined in an interval (a, b); in this case, the maximum SMI is obtained for the density function

$$f(x) = \frac{1}{b-a}, \quad \text{for } a \leq x \leq b, \tag{2.79}$$

and the corresponding value of the SMI is $H = \log(b-a)$.

The SMI associated with equilibrium density (2.78) is

$$\begin{aligned} H_{\text{max}} &= -\int_0^L f_{\text{eq}}(x) \log f_{\text{eq}}(x) dx \\ &= -\frac{1}{L} \log \frac{1}{L} \times \int_0^L dx = \log L. \end{aligned} \tag{2.80}$$

Clearly, the larger L is, the larger the SMI, or the larger the uncertainty in the location of a particle within the range $(0, L)$. Note carefully that L has units of length. Therefore, the only meaningful application of Eq. (2.80) is for *difference* in H. A few examples of the uniform distribution are shown in Fig. 2.11.

In anticipating the application of this result in Chap. 5, we divide the length L into n segments, each of length h, i.e. $L = h \cdot n$ (h will later be the Planck constant, but here it is an arbitrary unit of length); see Fig. 2.12. We can use the property of consistency to express

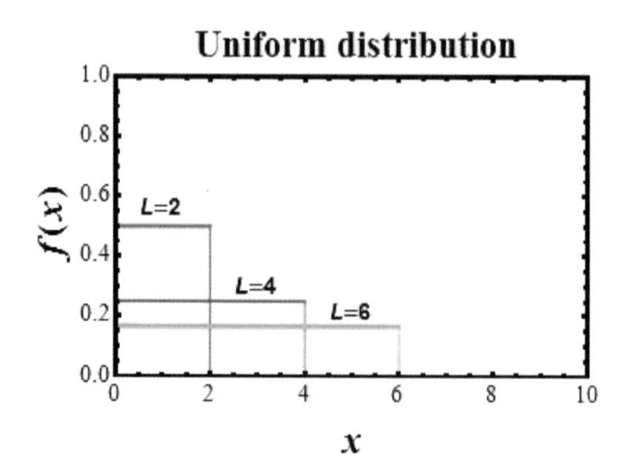

Fig. 2.11. The uniform distribution for different values of L.

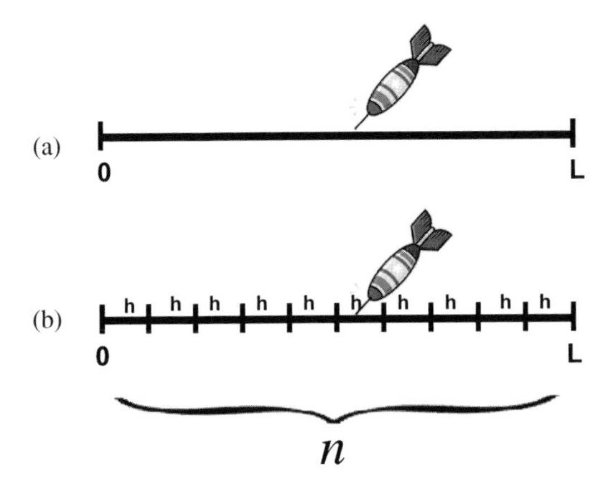

Fig. 2.12. (a) Transition from the continuous to the discrete case and (b).

$H(0, L)$ as

$$H(0,L) = H\left[\frac{1}{n}, \cdots, \frac{1}{n}\right] + \sum_{i=1}^{n} \frac{1}{n} H\left[h\right]. \tag{2.81}$$

This equation simply corresponds to rewriting the uncertainty regarding the location of the particle in the range $(0, L)$ in two terms; first, the uncertainty with respect to which of the n boxes of size h, and the average uncertainty in the location of the particle *within* the boxes of size h.

From Eqs. (2.80) and (2.81), we have

$$\begin{aligned} H(0,L) &= \log n + \sum_{i=1}^{n} \frac{1}{n} \log h \\ &= \log \frac{L}{h} + \log h = \log L, \end{aligned} \tag{2.82}$$

which is consistent with Eq. (2.80).

Now suppose that h is very small so that we do not care (or we cannot care) about the location within the box of size h. All we care about is in which of the n boxes the particle is located. Clearly, the uncertainty in this case is simply the SMI of a discrete and finite case, i.e.

$$H\left[\frac{1}{n},\cdots,\frac{1}{n}\right] = \log n = \log L - \log h. \tag{2.83}$$

Thus, the subtraction of $\log h$ from $\log L$ amounts to the passage from the continuous to the discrete case of determining in which of the n boxes the particle is located. In practice, we can never determine the location of a particle with absolute or infinite accuracy. There is always a short interval of length within which we cannot tell where the particle is. Hence, in all these cases, it is the discrete rather than the continuous definition of H that applies. Note also that if we do not care about the location within the small intervals of length h, we get a pure number under the logarithm in Eq. (2.83).

2.11.2 *The Normal Distribution*

The second theorem we present here is the following:

Of all the continuous distribution densities $f(x)$ for which the second moment (or the standard deviation) is finite and constant, the Gaussian (or normal) distribution maximizes the SMI defined in Eq. (2.71). The mathematical problem is to maximize H as defined in Eq. (2.71) subject to the two conditions

$$\int_{-\infty}^{\infty} f(x)dx = 1, \tag{2.84}$$

$$\int_{-\infty}^{\infty} x^2 f(x)dx = \sigma^2 \ (= \text{constant}). \tag{2.85}$$

We define the auxiliary functional

$$A\left[f(x)\right] = -\int f(x)\log[f(x)]dx + \lambda_1 \int f(x)x^2 dx + \lambda_2 \int f(x)dx. \tag{2.86}$$

The functional derivative of $A\left[f(x)\right]$ with respect to the component $f(x')$ is

$$\frac{\delta A}{\delta f\,(x')} = -1 - \log f(x') + \lambda_1 x'^2 + \lambda_2. \tag{2.87}$$

The condition for maximum H is thus

$$-1 - \log f^*(x) + \lambda_1 x^2 + \lambda_2 = 0. \tag{2.88}$$

The two Lagranges constants may be obtained by substituting Eq. (2.88) in Eqs. (2.84) and (2.85) to obtain

$$f^*(x) = \exp\left(\lambda_1 x^2 + \lambda_2 - 1\right) \tag{2.89}$$

$$= \int_{-\infty}^{\infty} f^*(x)dx = \exp(\lambda_2 - 1) \int_{-\infty}^{\infty} \exp\left(\lambda_1 x^2\right) dx$$

$$= \sqrt{-\frac{\pi}{\lambda_1}} \exp(\lambda_2 - 1), \tag{2.90}$$

$$\sigma^2 = \int_{-\infty}^{\infty} x^2 f^*(x)dx = \exp(\lambda_2 - 1) \int_{-\infty}^{\infty} x^2 \exp\left(\lambda_1 x^2\right) dx$$

$$= \sqrt{-\frac{\pi}{4\,(-\lambda_1)^3}} \exp(\lambda_2 - 1). \tag{2.91}$$

From the last two equations, we can solve for λ_1 and λ_2, to obtain

$$\lambda_1 = \frac{-1}{2\sigma^2}, \quad \exp(\lambda_2 - 1) = \frac{1}{\sqrt{2\pi\sigma^2}}. \tag{2.92}$$

Hence, the equilibrium density is

$$f^*(x) = f_{\text{eq}}(x) = \frac{\exp\left(-x^2/2\sigma^2\right)}{\sqrt{2\pi\sigma^2}}. \tag{2.93}$$

In Fig. 0.8, we showed a few examples of the normal distribution.

The maximum value of the SMI (note that we use the natural logarithm) is

$$H_{\max} = -\int_{-\infty}^{\infty} f^*(x)\log f^*(x)dx = \frac{1}{2}\log\left(2\pi e\sigma^2\right). \tag{2.94}$$

In the application of this result for the velocity distribution in one dimension (see Chap. 5), we have the probability distribution density for

$$f^*(v_x) = \sqrt{\frac{m}{2\pi k_B T}}\exp\left(\frac{-mv_x^2}{2k_B T}\right), \tag{2.95}$$

where we identify the standard deviation σ^2 as [see Chap. 5 and Ben-Naim (2008)]

$$\sigma^2 = \frac{k_B T}{m}, \tag{2.96}$$

where k_B is the Boltzmann constant and T the absolute temperature. The SMI associated with the velocity distribution is thus

$$H(v_x) = \frac{1}{2}\log\left(2\pi e k_B T/m\right). \tag{2.97}$$

Similarly, for the momentum distribution in one dimension $p_x = mv_x$, we find that the density distribution which maximizes H is

$$f^*(p_x) = \frac{1}{\sqrt{2\pi m k_B T}}\exp\left(\frac{-p_x^2}{2mk_B T}\right) \tag{2.98}$$

and the corresponding maximum value of the SMI is

$$H_{\max}\left(p_x\right) = \frac{1}{2}\log\left(2\pi e m k_B T\right). \tag{2.99}$$

In Chap. 5, we will use this last expression to construct the analog of the Sackur–Tetrode equation for the entropy of an ideal gas.

The significance of the last result is that σ^2 is proportional to the temperature of the gas, and the temperature of the gas is related to the average kinetic energy of the particles in the gas. Hence, the last result means that given a temperature or, equivalently, fixing the total kinetic energy of the gas molecules, the Gaussian or normal distribution of velocities v_x, v_y, and v_z produces the maximum value of the SMI. This is the equilibrium Maxwell–Boltzmann distribution of velocities (see Chap. 5).

2.11.3 *The Exponential or Boltzmann Distribution*

In this case, we want to find the maximum of the H defined in Eq. (2.71) subject to the two conditions

$$\int_0^\infty f(x)dx = 1, \tag{2.100}$$

$$\int_0^\infty xf(x)\,dx = a, \quad \text{with a} > 0. \tag{2.101}$$

Using the Lagrange multipliers λ_1 and λ_2, we seek the maximum of the functional

$$A\left[f(x)\right] = -\int_0^\infty f(x)\log f(x)dx + \lambda_1 \times \int_0^\infty f(x)dx + \lambda_2 \int_0^\infty xf(x)dx. \tag{2.102}$$

The condition for an extremum is obtained by taking the functional derivative of $A\left[f(x)\right]$, and setting it equal to zero, i.e.

$$-1 - \log f^*(x) + \lambda_1 + \lambda_2 x = 0, \tag{2.103}$$

or, equivalently,

$$f^*(x) = \exp\left(\lambda_2 x + \lambda_1 - 1\right). \tag{2.104}$$

Substituting this density function in the two constraints (2.100) and (2.101), we get

$$\exp\left(\lambda_1 - 1\right)\int_0^\infty \exp\left(\lambda_2 x\right)dx = 1, \tag{2.105}$$

$$\exp(\lambda_1-1)\int_0^\infty x\exp(\lambda_2 x)dx = a. \tag{2.106}$$

Note that λ_2 cannot be positive, otherwise the two constraints cannot be satisfied (not even for $\lambda_2 = 0$). From Eqs. (2.105) and (2.106), we obtain

$$\exp(\lambda_1-1)\left[\frac{\exp(\lambda_2 x)}{\lambda_2}\right]_0^\infty = -\frac{\exp(\lambda_1-1)}{\lambda_2} = 1, \tag{2.107}$$

$$\exp(\lambda_1-1)\left[\frac{(\lambda_2 x - 1)\exp(\lambda_2 x)}{\lambda_2^2}\right]_0^\infty = \frac{\exp(\lambda_1-1)}{\lambda_2^2} = a. \tag{2.108}$$

These two equations can be solved for λ_1 and λ_2 to obtain

$$\lambda_2 = \frac{-1}{a}, \tag{2.109}$$

$$\exp(\lambda_1-1) = \frac{1}{a}. \tag{2.110}$$

Hence, the density function that maximized H is

$$f^*(x) = \frac{1}{a}\exp\left(\frac{-x}{a}\right). \tag{2.111}$$

In Fig. 2.13, we show a few examples of the exponential distribution for different values of a.

The value of the SMI corresponding to this density is (note that log is the natural logarithm)

$$H_{\max} = \log a + 1 = \log(ae). \tag{2.112}$$

2.12 Examples

We conclude this chapter with a few examples of experiments having well-defined probability distributions and hence also well-defined SMI. These examples will be of use to us later on in this book, when we

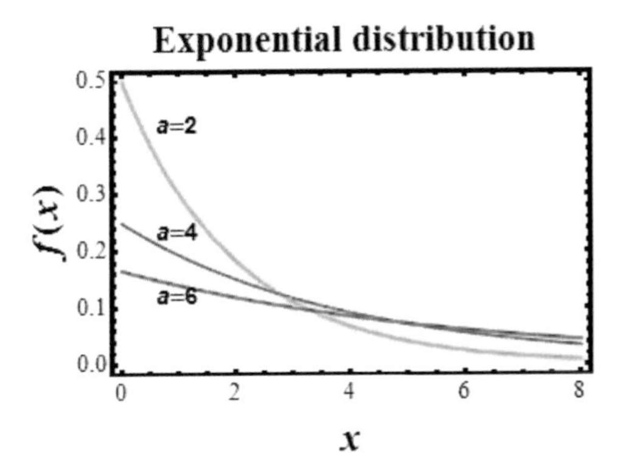

Fig. 2.13. The exponential distribution for different *a*'s.

study dependence between experiments, multivariate mutual information, and Markov chains. The reader is urged to study these examples carefully.

2.12.1 *Spins Having Two Possible Orientations: "Up" and "Down"*

Consider an ideal gas of molecules, each having a dipole moment. The dipole could be electric (with + and – charges) or magnetic (with "north" and "south" poles). We will refer to these particles as *spins*, and we assume that each spin can be in either of the two "states" referred to as "up" and "down." We also assume that the particles are independent in the sense that the *state* of one particle has no effect on the probability of other particles to be in any specific state. In addition, we assume that there is an external field denoted as F which affects the state of each particle — for instance, an electric field which favors, say, the orientation of the electric dipole moment, in the "up" state; see Fig. 2.14(a).

The probabilities of the different states of the spins are derived from statistical mechanics. In this chapter, we will accept these probabilities for calculating the SMI.

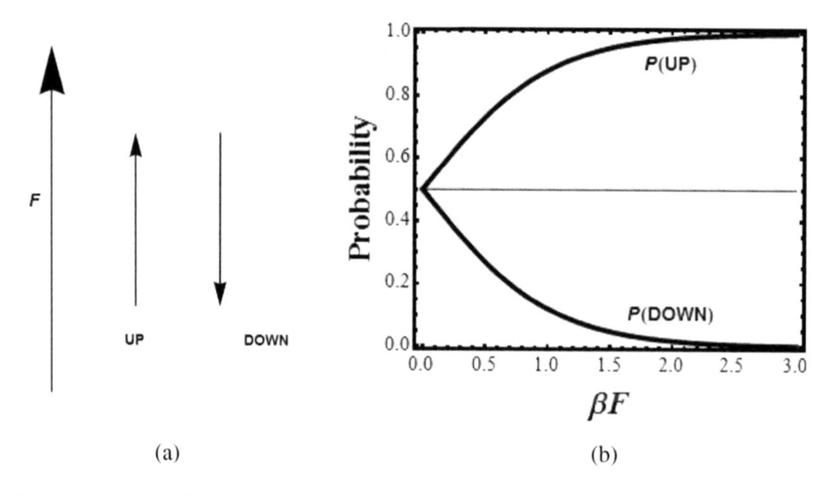

Fig. 2.14. (a) The two orientations of the spin in an external field. (b) The probability of "up" and "down" as a function of βF.

Some justification for the choice of these probabilities will be discussed in Chap. 5 [and for more details the reader is referred to any textbook on statistical mechanics — say, Ben-Naim (1992)].

Thus, we assume that the probability associated with the *state x* is given by

$$p(x) = \frac{\exp\left[-\beta U(x)\right]}{Z_1}, \tag{2.113}$$

where $U(x)$ is the energy of the spin. $\beta = T^{-1}$, with T the absolute temperature, and z_1 is the normalization constant.

In this example, we have two states only. We will assign the numerical values of $+1$ and -1 to the two states "up" and "down," respectively.

Thus, each spin has two energy states,

$$\begin{aligned} U(x=1) &= -Fx = -F, \\ U(x=-1) &= -Fx = +F, \end{aligned} \tag{2.114}$$

and the corresponding probabilities

$$p(x=1) = \frac{\exp(+\beta F)}{z_1},$$
$$p(x=-1) = \frac{\exp(-\beta F)}{Z_1}, \tag{2.115}$$

with the normalization constant

$$Z_1 = \sum_{x=\pm 1} \exp(-\beta F x). \tag{2.116}$$

If we choose F to be positive, then the "up" state (i.e. $x = 1$) will be the preferred state of the spin, i.e.

$$p(x=1) \geq p(x=-1). \tag{2.117}$$

The equality sign holds when either $F = 0$ (i.e. no external field) or $T \to \infty$ (infinite temperature).

Clearly, when $F = 0$, there is no preference for any state, and the two states are equally probable, i.e. in this case

$$p(x=1) = p(x=-1) = \frac{1}{2} \tag{2.118}$$

For any finite F, the state "up" is preferred. Figure 2.14(b) shows the probabilities $p(1)$ and $p(-1)$ as a function of βF. We see that when $F = 0$ (or, equivalently, $T = \infty$), we have the equality (2.118). As βF increases, the probability of the "up" state increases from 1/2 to 1, while the probability of the "down" state decreases from 1/2 to 0.

The reader is urged to pause and examine the physical meaning of the two curves in Fig. 2.14(b). Suppose that $F > 0$ is fixed, and we change the temperature from zero to infinity, or fix $T > 0$ and change the field, how would you expect the probabilities to change?

Figure 2.15(a) shows the SMI of the system per spin as a function of T for a fixed value of $F = 1$. As expected, at $\beta F = 0$ (either zero field or infinite temperature), the value of the SMI is 1, indicating total randomness, i.e. equal probability for the two states of the spin.

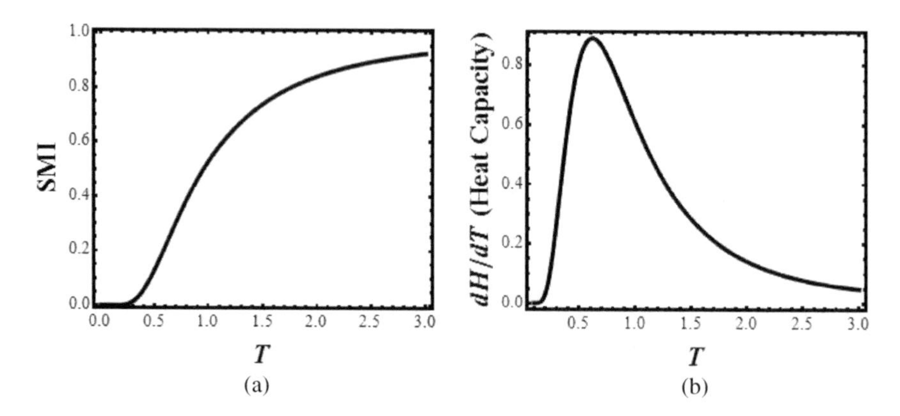

Fig. 2.15. The SMI as a function of T, and its derivative with respect to T.

At very low temperatures (either $F \to \infty$ or $T \to 0$), the spins will preferentially orient themselves in the direction of the field. In this case the SMI tends to zero.

For any finite field F, we have two competing effects: the field, which tends to orient the spins in a preferred direction, and the temperature (i.e. the kinetic energies of the particles), which tends to randomize the orientations of the spin. Thus, at very high temperatures, the temperature effect wins the competition, leading to complete randomness. Again, the reader is urged to pause and examine the physical significance of the curve in Fig. 2.15(a).

In Fig. 2.15 we drew the SMI as a function of T for a fixed F (i.e. instead of $\beta F = F/T$, we fixed $F = 1$, and drew the SMI as a function of T). This is the more "conventional" view of the entropy as a function of the temperature. This function is always a monotonic increasing function of T (see also Chap. 5). Note, however, that it starts at $T = 0$ with SMI $= 0$, and with a zero slope. As T increases, the slope initially increases, then decreases. This is a typical behavior of the heat capacity of a system of particles having two states. We show in Fig. 2.15(b) the analog of the *heat capacity* for this system, i.e. the derivative of the SMI with respect to the temperature.

2.12.2 *Spins Having Multiple Orientations*

We now extend the model of the previous section. Again, we have a system of noninteracting spins in an external field F. Instead of only two possible orientations, we allow n orientations for each spin. Each orientation is specified by the angle α with respect to the direction of the field F. Thus, the orientation $\alpha = 0$ is the most favored, one in the presence of the field, and $\alpha = \pi$ is the most unfavored orientation. Figure 2.16 shows the case of seven orientations, i.e. $n = 7$ with

$$\alpha = 0, \quad \frac{\pi}{6}, \quad \frac{2\pi}{6}, \quad \frac{3\pi}{6}, \quad \frac{4\pi}{6}, \quad \frac{5\pi}{6}, \quad \frac{6\pi}{6} = \pi. \qquad (2.119)$$

Figure 2.17 shows the probabilities of finding the spins in different orientations for the case $n = 10$. As we can see, all the curves start at a value of 0.1 at $\beta F = 0$ (either $F = 0$ or $T \to \infty$), which means total randomness, or equal probability for each orientation.

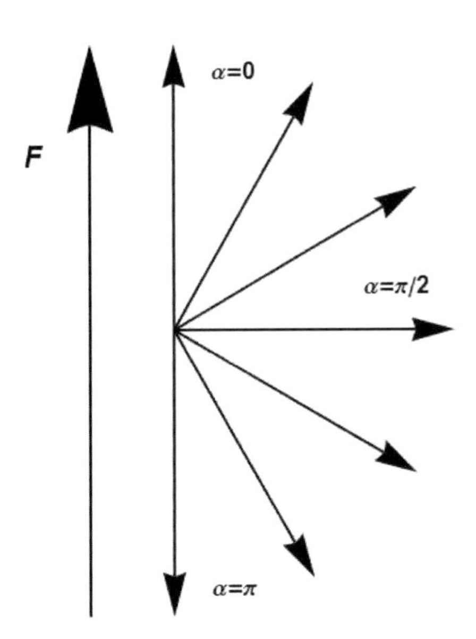

Fig. 2.16. A spin with a finite number of possible orientations; see Eq. (2.119).

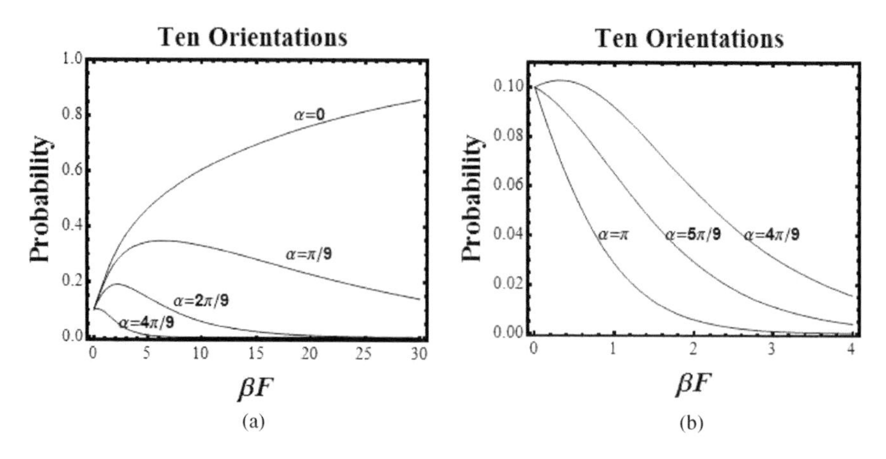

Fig. 2.17. The probabilities of the various orientations for the system of spins with a finite number of possible orientations.

Since the $\alpha = 0$ orientation is favored by the field, it is clear that when βF *increases* (i.e. either T fixed and F increases, or F fixed and T decreases), the probability of finding this orientation will *increase* and eventually reach the value of 1 at $\beta F \to \infty$. On the other hand, the orientation $\alpha = \pi$, the least favored, will *decrease* as βF increases; see Fig. 2.17(b).

An interesting behavior is found for $0 < \alpha < \pi/2$. All these orientations [in our particular example, they are $\alpha = 2\pi/9, 3\pi/9, 4\pi/9$; only these are shown in Fig. 2.17(a)] are favored by the field. Therefore, starting with probability 0.1 at $\beta F = 0$, the probabilities of all of these orientations initially increase with increasing βF. However, since eventually the probability of the orientation $\alpha = 0$ tends to 1, as βF tends to infinity, the probabilities of all these orientations will start to decrease at large values of βF. As we can see, all the curves for $\alpha < \pi/2$ (excluding $\alpha = 0$) will tend to 0 at $\beta F \to \infty$.

Thus, the fact that in the limit $\beta F \to \infty$ the probability of one orientation, $\alpha = 0$, tends to 1 affects all the other probabilities. This is so since the sum of the probabilities must be 1. Another way to see this behavior is to note that initially, when we "turn on" the field

($\beta F \neq 0$), both $\alpha = 0$ and $\alpha = \pi/9$ will be favored by the field, and their probabilities will *increase* as βF increases. However, as we can see from Fig. 2.17(a), the *rate of increase* of the probability of $\alpha = 0$ is larger than the rate for $\alpha = \pi/9$ (i.e. the initial slope of the curve for $\alpha = 0$ is larger than that of $\alpha = \pi/9$). Since the sum of all probabilities must be 1, the rate of increase of the probability $\alpha = 0$ must cause an eventual decrease in the probabilities of all other orientations.

It should be noted that the behavior of the $\alpha = 0$ case is not specific to the orientation in the direction of the field. For any other set of orientations, the one most favored by the field will always "grab" the total probability as $\beta F \to \infty$. In the following two exercises, we do not include the case $\alpha = 0$, and yet the general behavior of the system is the same.

Exercise 2.7

Repeat the calculations of the probabilities and the SMI for the case of five orientations: $\alpha = \pi/9, 2\pi/9, 3\pi/9, 4\pi/9$, and $5\pi/9$.

Exercise 2.8

Repeat the calculations as in Exercise 2.7, for the case of three orientations: $\alpha = \pi/3, \pi/2$, and $2\pi/3$.

Again, the reader is urged to pause and think about the two competing effects of the field and the temperature which produce the behavior of the curves in Fig. 2.17. This understanding is essential for the understanding the various SMI's calculated for this system.

A different behavior is shown for $\alpha > \pi/2$ in Fig. 2.17(b). Only two are shown in the figure. We see that the probabilities of these orientations *decrease monotonically* with βF.

Figure 2.18 shows the SMI as a function of βF for $n = 6, 10, 20$. We see that the SMI starts with the value of $\log_2 n$ (i.e. the total randomness) at $\beta F = 0$ (either $F = 0$ or $T = \infty$). Here the values are SMI $= 2.58, 3.32, 4.32$ for $n = 6, 10, 20$, respectively. The value of the SMI decreases monotonically as βF increases. Eventually, all the

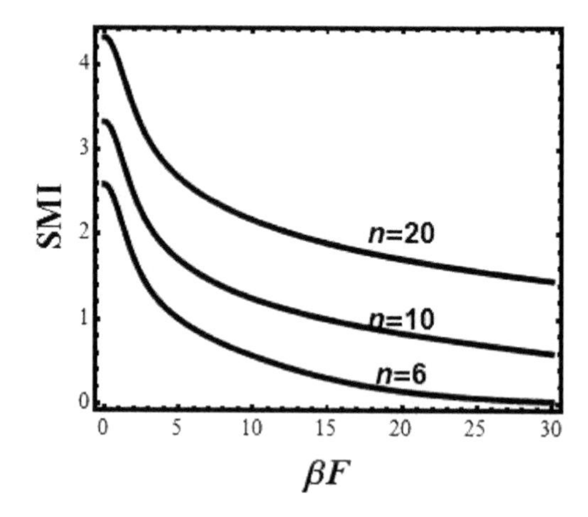

Fig. 2.18. The SMI of the system of spins with different finite numbers, n, of possible orientations.

curves tend to zero, meaning that one orientation becomes certain. The reader is urged to pause and examine the curves in Fig. 2.18. Explain to yourself why each curve decreases with βF. Why is the entire curve for a larger n above the curve for a smaller n?

2.12.3 *The SMI of the Letters of an Alphabet*

Perhaps the most important probability distribution relevant to IT is the frequencies of the letters of the alphabet in a given language.

We can view any message sent through a communication channel as a flow of sequence of symbols. Each symbol can be in one of the "states" which are the letters of the alphabet in a particular language.

If all the letters of the alphabet are equally probable, then the corresponding SMI per letter will be $\log_2 n$, where n is the number of letters in the alphabet. In reality, the frequencies of the letters of an alphabet are very different from each other. Figure 2.19 shows the frequencies of the various letters in English (sometimes one adds the "space" as an additional "letter" in the counting of the size of the alphabet). These frequencies may be viewed as the *empirical probabilities* of the various letters. Of course, they may be different in

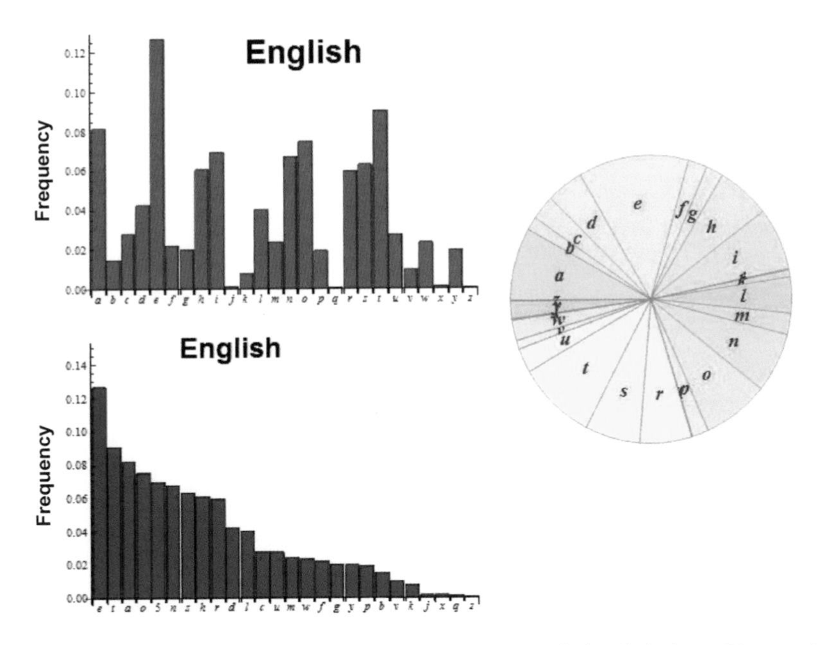

Fig. 2.19. The distribution of letters in the English alphabet. Upper, in alphabetical order; lower and right, ordered by frequency.

different languages, using the same letters (say, French and Italian), or even in different pieces of literature in the same language. An interesting example is given below.

As can be seen from Fig. 2.19(b), the letter "e" is by far the most frequently used letter. The vowel "a" is the third after "t". If you are offered the challenge of guessing a letter in English by asking binary questions, it will be convenient to use the pie chart in Fig. 2.19(c). This will allow you to divide the entire alphabet into two groups of almost equal probabilities.

The largest frequency of the letter "e" is also true of other languages using the same alphabet such as French, Dutch, and German; see Fig. 2.20.

In Italian, the letters "a" and "e" have nearly the same frequencies; see Fig. 2.20(b). However, in Portuguese, Turkish, and Polish, the letter "a" is the most frequent. Note that in Polish the letter "z" has

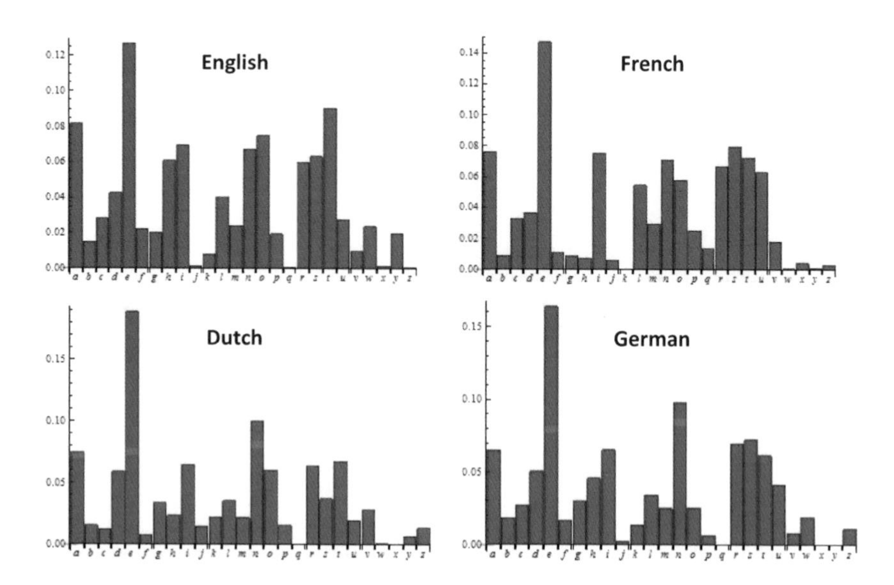

Fig. 2.20(a). The distribution of letters in different languages.

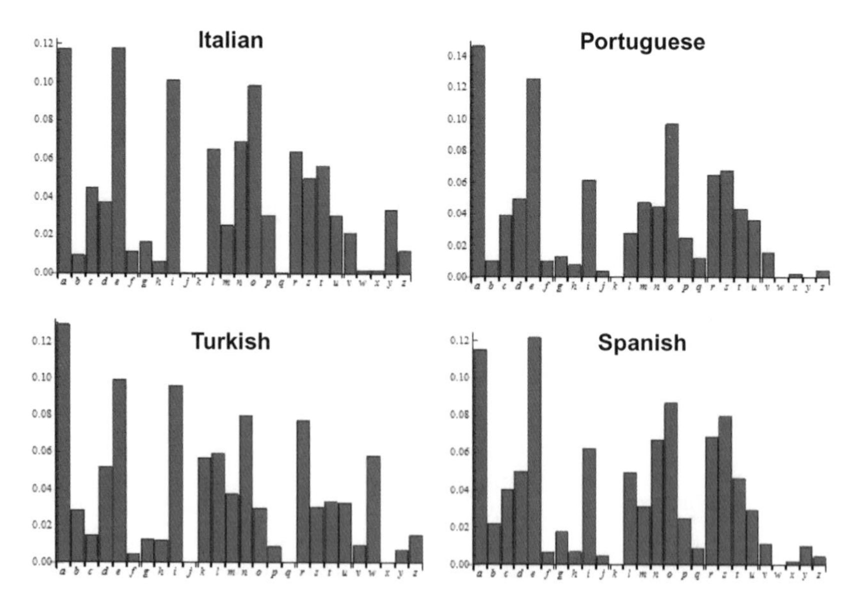

Fig. 2.20(b). The distribution of letters in different languages.

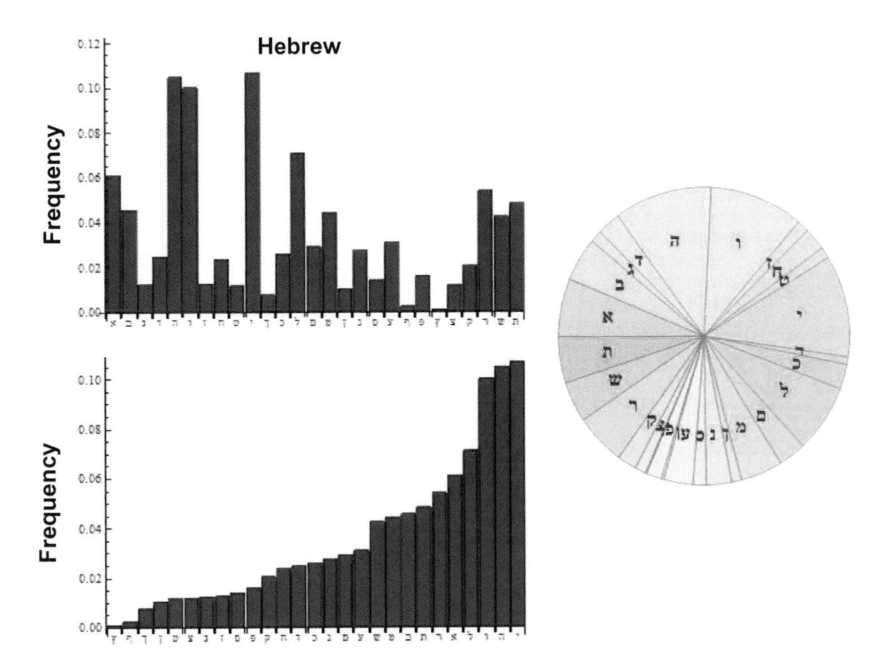

Fig. 2.21. The distribution of letters in the Hebrew alphabet. Upper, in alphabetical order; lower and right, ordered by frequency.

quite a big frequency compared with other languages. (Note also that in these figures we drew only frequencies of the letters "a" to "z" as in English, not including special symbols of the language.) Figure 2.21 shows the frequencies of the Hebrew alphabet. Although there are no vowels in the Hebrew alphabet as in other languages, some letters serve occasionally as vowels, and have the highest frequencies.

2.13 Redundancy

2.13.1 *Qualitative Meaning*

Another concept that is used in IT is the *redundancy* of a message. As always in IT, one must clearly distinguish between the qualitative, colloquial meaning of a concept and the corresponding meaning of the theoretical concept.

We start by introducing the concept of redundancy in a qualitative manner. At one TV station, the weather forecast for the next week was: "It will rain every day in the coming week." At another TV station, the same forecast was stated as:

"It will rain on Sunday, it will rain on Monday, it will rain on Tuesday, it will rain on Wednesday, it will rain on Thursday, it will rain on Friday, and it will rain on Saturday."

Obviously, the two forecasts provided the *same information*. It is clear that the specification of each of the days in the second was redundant.

Redundancy is sometimes used in daily life to avoid any misunderstanding. Sometimes it involves the repetition of the same message again and again. In other cases one repeats the same message using slightly different words but the overall message is the same. The famous biblical story of Jacob and Laban is well-known.

Jacob met Rachel near a well and fell in love with her. He then went to her father, Laban, who happened to have another daughter who was older than Rachel, named Leah.

Jacob said to Laban: "I will serve you seven years for Rachel, your younger daughter." (Genesis 29:18)

It is believed that Jacob wanted to secure accuracy for his message. Obviously, Laban knew that Rachel was his younger daughter, and yet Jacob added the redundant information "your younger daughter."

Today, the phrase "Rachel, your younger daughter" is used as an idiomatic expression in Hebrew, especially among lawyers who want to remove any doubts or ambiguity when they invoke this phrase to emphasize the meaning of their statement. Of course, everyone notices that there is redundancy in this phrase. Laban knew that Jacob was in love with Rachel, and there was no need to add that Jacob was referring to Laban's *daughter*, and that he was referring to Laban's *younger* daughter. Perhaps Jacob knew about the dubious character of Laban and in order to avoid any misunderstanding he added "your younger daughter" to ensure that Laban would get the right message and could not claim any misunderstanding.

The irony of the story is that in spite of Jacob's unambiguous message, Laban did cheat Jacob and, under the cover of darkness, he gave him his older daughter, Leah, instead of Rachel.

Sometimes, the whole information given is redundant. This is when everyone already knows that information.

Hofstadter (1985) quoted from some poems which were intentionally written to be nonsensical poems. It is clear that the ultimate redundant message is a nonsensical message — or a message that conveys no information at all. As we will see below, even a completely redundant nonsensical message can be assigned a measure of information. This redundancy has nothing to do with the redundancy of information used in daily life. On the one hand, a message can be highly redundant in a colloquial sense, but not have high redundancy in IT. On the other hand, a message can have zero redundancy in a colloquial sense, but high redundancy in the informational-theoretical sense.

2.13.2 *Redundancy Defined in Terms of SMI*

The term "redundancy" used colloquially means that something is excessive, superfluous, or repetitious, or excessive wordiness, text, etc. For more details, see BenNaim (2015a, 2016a). In IT, this term was distilled from the general term and is defined in a more precise manner. For any experiment with a given distribution of the outcomes $p_1, \ldots, p_N$, we can define the fraction

$$\frac{\mathrm{SMI}\left(p_1, \ldots, p_N\right)}{\mathrm{Max\,SMI}} = \frac{\mathrm{SMI}}{\log_2 N}. \tag{2.120}$$

We know that the maximum SMI for an experiment with a given N is obtained when all the outcomes are equally likely, i.e. their probabilities are $p_i = 1/N$ (for $i = 1, 2, \cdots N$). Therefore, the maximum value of the SMI is simply $\log_2 N$.

In Fig. 2.22 we show three games, each with four outcomes. The corresponding SMI's are SMI(a) = 2, SMI(b) = 179, and SMI(c) = 121. These values indicate that it is "easiest" to play game (c), in

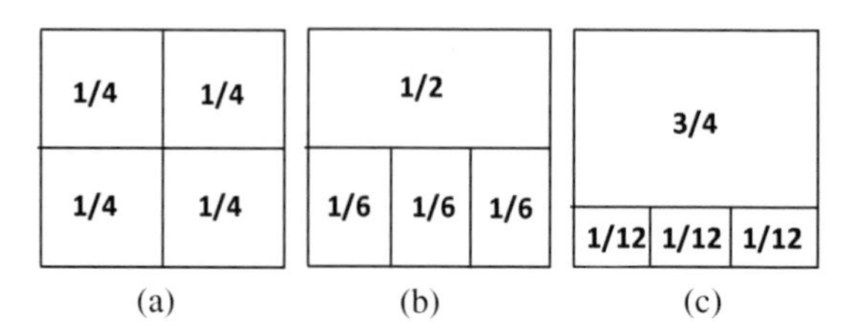

Fig. 2.22. Three games each with four possibilities, but with different probability distributions.

the sense that the distribution in (c) provides more information, and therefore one needs to ask fewer questions. On the other hand, we can say that the game on the left (a) does not provide any *information* we can use to our advantage (i.e. reducing the number of questions).

The redundancy as defined in IT is a measure of how much a game is easier to play relative to the one with the same number of outcomes but with equal probabilities. The formal definition is

$$R = \frac{\log_2 N - \text{SMI}}{\log_2 N} = 1 - \frac{\text{SMI}}{\log_2 N}. \tag{2.121}$$

For the three games in Fig. 2.22, we have $R\,(\text{a}) = 0$, $R\,(\text{b}) = 0.104$, $R\,(\text{c}) = 0.39$. This means that the distribution in (c) is the most redundant. This is vaguely reminiscent of the redundancy in spoken language.

The concept of redundancy is useful in the application of IT to the transmission of information. The redundancy in this case is a measure of how easy it is to guess a letter from an alphabet when we know the frequency of its occurrence in that particular language.

2.13.3 *Redundancy in the Letters of an Alphabet*

The following data is taken from Yaglom and Yaglom (1983):

For the English language, the maximum SMI is

$$H_0 = \log 27 \approx 4.75 \text{ bits} \tag{2.122}$$

(H_0 refers to the case where all symbols are equally probable).

Fig. 2.23. The cover of the book Gadsby by Ernest Vincent Wright.

Taking the real frequencies of letters in the English language, one gets

$$H_1 = -\sum p_i \log p_i \approx 403\,\text{bits}, \tag{2.123}$$

where p_i is the relative frequency of the letter i.

If we know the distribution of pairs of letters, we can define the *conditional* SMI of a letter i followed by a letter j

$$H_2 = -\sum p(i\,|\,j)\log[p(i\,|\,j)].$$

(We will discuss conditional SMI in the Sec. 3.1. Here, we use the definition of SMI applied to the conditional probabilities. It is the difference between the SMI defined on pairs of two consecutive letters, minus the SMI of single letters of the alphabet.)

Table 2.1

Language	English	French	Italian	Portuguese	Russian
H_0	4.75	4.75	4.39	4.52	5.00
H_1	4.03	3.95	3.90	3.91	4.35
H_2	3.32	3.17	3.32	3.35	3.52
H_3	3.10	2.83	2.76	3.20	3.01

One can continue to define the conditional information of a letter k followed by two letters i and j, by

$$H_3 = -\sum p(k|i,j)\log[p(k|i,j)].$$

Table 2.1 shows the values of the SMI and the conditional information for single letters in different languages.

As expected, we always have $H_0 \geq H_1$, and the value of the conditional information decreases from H_1 to H_2 to H_3, etc.

It is interesting to note that in English (as well as other European languages) the vowels are more frequent than the consonants. In modern day Hebrew, the vowels are not used in written text. This fact dramatically changes the redundancy in the Hebrew language. As an example, Bluhme [quoted in Yaglom and Yaglom (1983)] compared the statistical characteristic of a collection of three-letter words in Hebrew and in English, and found the following results:

$$H_3^{(\mathrm{Heb})} \approx 3.73\left(\frac{\mathrm{bit}}{\mathrm{letter}}\right),$$

$$R_3^{(\mathrm{Heb})} = 1 - \frac{H_3}{H_0} \approx 0.16,$$

$$H_3^{(\mathrm{Eng})} \approx 0.83\left(\frac{\mathrm{bit}}{\mathrm{letter}}\right),$$

$$R_3^{(\mathrm{Eng})} \approx 0.82.$$

Note that the redundancy (for the three-letter words) in Hebrew is significantly smaller than in English.

Similar analysis of H_0, H_1, and H_2 was done in the arts, and in many other fields.

For an interesting discussion on the SMI of different languages, as well as of different books by the same author (e.g. Shakespeare), see Landsberg (1961).

We have discussed the term "redundancy" in its colloquial sense, and its definition within IT. As we have seen, these two concepts of redundancy are very different; one is qualitative and the other is quantitative.

Confusing the two is very common in the popular science literature. For some examples, see Ben-Naim (2015a, 2016a).

2.13.4 *Wright's Novel Without the Letter "e"*

As we have noted above, the letter "e" is the most frequently used letter in English. This does not necessarily mean that the frequency of occurrence of "e," or any other letter, is the same in different documents. In 1939, Ernest Vincent Wright published a novel of over 50,000 words titled *Gadsby*, which does not include the letter "e", Fig. 2.23. In the introduction the author explains how he did it:

From the introduction to Gadspy by Ernest Vincent Wright:

> "THE ENTIRE MANUSCRIPT of this story was written with the E type-bar of the typewriter tied down; thus making it impossible for that letter to be printed. This was done so that none of that vowel might slip in, accidentally; and many did try to do so!
>
> I have even ordered the printer not to head each chapter with the words "Chapter 2," etc., on account of that bothersome E in that word."

Here are a few paragraphs from Chapter 1:

> "If youth, throughout all history, had had a champion to stand up for it; to show a doubting world that a child can think; and, possibly,

do it practically; you wouldn't constantly run across folks today who claim that "a child don't know anything."

A child's brain starts functioning at birth; and has, amongst its many infant convolutions, thousands of dormant atoms, into which God has put a mystic possibility for noticing an adult's act, and figuring out its purport. Up to about its primary school days a child thinks, naturally, only of play. But many a form of play contains disciplinary factors. "You can't do this," or "that puts you out," shows a child that it must think, practically, or fail. Now, if, throughout childhood, a brain has no opposition, it is plain that it will attain a position of "status quo," as with our ordinary animals. Man knows not why a cow, dog or lion was not born with a brain on a par with ours; why such animals cannot add, subtract, or obtain from books and schooling, that paramount position which Man holds today. But a human brain is not in that class. Constantly throbbing and pulsating, it rapidly forms opinions; attaining an ability of its own; a fact which is startlingly shown by an occasional child "prodigy" in music or school work. And, as with our dumb animals, a child's inability convincingly to impart its thoughts to us, should not class it as ignorant."

Upon this basis I am going to show you how a bunch of bright young folks did find a champion; a man with boys and girls of his own; a man of so dominating and happy individuality that Youth is drawn to him as is a fly to a sugar bowl. It is a story about a small town . . . I shall act as a sort of historian for this small town:

So, now to start our story. Branton Hills was a small town in a rich agricultural district; and having many a possibility for growth. But, through a sort of smug satisfaction with conditions of long ago, had no thought of improving such important adjuncts as roads; putting up public buildings, nor laying out parks; in fact a dormant, slowly dying community. So satisfactory was its status that it had no form of transportation to surrounding towns but by railroad, or "old Dobbin."

And the end:

> "Full moon sails across a sky without a cloud. A crisp night air has folks turning up coat collars and kids hopping up and down for warmth. And that giant star, Sirius, winking slyly, knows that soon, now, that light up in His Honor's room window will go out. A glorious Fttt! It is out! So, as Sirius and Luna hold an all-night vigil, I'll say a soft "Good-night" to all our happy bunch, and to John Gadsby — Youth's Champion." FINIS.

Obviously, it is very easy to write a paragraph or a whole book avoiding the least frequent letters such as "z", "q", "j" or "x" (such a text is called a *lipogram*, which means "leaving out a letter"). It is far more challenging to avoid a very frequently used letter, such as "e" or "a".

Exercise 2.9

Can you estimate the SMI associated with this text, assuming that the letter "e" does not appear in the entire text but the relative frequencies of all other letters are the same as in English? Is the assumption made above plausible? Why might the elimination of one letter change the distribution of all other letters?

2.13.5 *An Amusing Example Involving Redundancy*

Finally, I would like to conclude this section with a very interesting example, which I believe is related to the redundancy in the English language.

Start by reading the following paragraph as quickly as you can:

> You arne't ginog to blveiee that you can aulactly uesdnatnrd what I am wirtnig. Beuacse of the phaonmneal pweor of the hmuan mnid, aoccdrnig to a rscheearch at Cmabrigde Uinervtisy, it deosn't mttaer in what order the ltteers in a wrod are, the olny iprmoatnt tihng is

> taht the frist and lsat ltteer be in the rghit pclae. The rset can be a taotl mses and you can sitll raed it wouthit a porbelm. Tihs is bcuseae the huamn mnid deos not raed ervey lteter by istlef, but the word as a wlohe. Amzanig huh? Yaeh and you awlyas tghuhot slpeling was ipmorantt!

This quotation was taken from Patel (2008), based on Rawlinson's thesis (1976) on "The Significance of Letter Position in Word Recognition." Patel used this example in connection with the "language" of the proteins written in amino acids (see Chap. 3). The explanation as to why one can read and understand the paragraph in spite of the many "spelling mistakes" is that we tend to read *whole words*, and not letter by letter. (Note that from the point of IT this paragraph has the same SMI as the original, correct paragraph.)

In my opinion, the main reason for recognizing the words in spite of the jumbled letters is the redundancy in the English language. Although I cannot offer a proof of my contention, I can offer a plausible argument. As noted above, the Hebrew language is far less redundant than the English language (this is due mainly to the absence of most vowels in written Hebrew).

If my conjecture is correct, then I would expect that the less the redundancy is in the language, the more difficult it would be to read text with jumbled letters. Indeed, I tried several examples and found to my surprise that even when I *knew* the original text of the paragraph (because I had *invented* it), I had a hard time reading the jumbled letters. Here is an example of two sentences in Hebrew. If you know Hebrew, try to read this as quickly as you can (the original sentences are given at the end of this section).

בירעבת כובתה ברדך כלל חורסת הועתונת
למיפעם זה משקה על הויהגי הוכנן של הילמים.

לורמת הועדבה שנאי ששיתבי את היותאות בצמעי,
לא יותלכי לרוקא את השפטמים השושבמים.

I also notice that because of the lack of vowels, many words in the jumbled letters have a meaning in Hebrew, differing from the original words. Another explanation for this phenomenon was offered by Velan and Frost (2007). They presented sentences in both English and Hebrew to bilingual students. It was found that word recognition was much more difficult for jumbled Hebrew letters than for English. These authors pointed out that although the English reader tends to read *whole words*, the Hebrew reader first tries to identify the *triconsonantal roots* in the Hebrew words. This identification is sometimes difficult, because the same three consonants usually have different meanings when written in a different order.

This fact makes it very difficult to read quickly the jumbled letters text in Hebrew. Hence, the authors concluded: it seems that *research at Hebrew University* may produce quite different results than *rsheearch at Cmabrigde Uinervtisy*, where visual word recognition is concerned.

Here are the correct sentences for the example presented above:

בעברית כתובה בדרך כלל חסרות התנועות
לפעמים זה מקשה על ההיגוי הנכון של המילים.

למרות העובדה שאני שיבשתי את האותיות בעצמי,
לא יכולתי לקרוא את המשפטים המשובשים.

2.14 Summary of Chapter 2

In this chapter, we have presented the basic properties of SMI. We found that the experiment having two outcomes leads to the definition of the unit of information: the bit. This definition of the bit is different from that of the binary digit used in computer science.

We then discussed the general case of a game or an experiment having n outcomes. We saw the relationship between the SMI and the number of binary questions one needs to ask in order to find a selected object out of a total of n objects. We also introduced the SMI defined on a continuous distribution, and derived the three fundamental distributions: uniform, normal, and exponential.

3

Conditional and Mutual Information

In this chapter, we introduce two important and useful concepts. To do that, we start with two experiments (or two rv's), X and Y, and we define the *joint* SMI, the conditional SMI (known as conditional information, or conditional entropy), and the mutual information. In Chap. 4, we will generalize these quantities to more than two experiments.

As for notation, we always assume that an experiment X (or a game, an rv, or a partition), having n outcomes, say $x_1, x_2, \ldots, x_n$, and a probability distribution $p_1, \ldots, p_n$. p_i is the probability of occurrence of the ith event. If X is an rv, then p_i is the probability that the event $(X = x_i)$ has occurred. We will use the notation p_i or $p(i)$ whenever it is clear which the rv or the experiment is. When we have two or more rv's, we need to specify which rv we are referring to. For instance, we will use the notation $p_X(i) = p(X = x_i)$ and $p_Y(i) = p(Y = y_i)$ for the probability of the ith outcome of the rv's X and Y, respectively.

It is sometimes advantageous to use the notation $p(x) = p(X = x)$ for the probability of the event $X = x$, i.e. the outcome of the experiment X is x, or the value of the rv X is x. x is sometimes referred to as a "letter" in a given alphabet.

3.1 Conditional Information

We start with the SMI defined on the joint probabilities $p(x, y) = p(X = x, Y = y)$, i.e. the probability of the event $X = x$ *and* the event $Y = y$. The SMI for this distribution is

$$H(X, Y) = \sum_{x,y} p(x, y) \log p(x, y), \tag{3.1}$$

where the summation is over all possible values of $x$and y.

We define the marginal probabilities

$$p(x) = \sum_{y} p(x, y), \tag{3.2}$$

$$p(y) = \sum_{x} p(x, y). \tag{3.3}$$

It should be clear that $p(x)$ is the probability of the event $X = x$, and $p(y)$ is the probability of the event $\mathbf{Y} = y$. Note again that the notation $H(x)$ means that the SMI is associated with the experiment, or the rv X. The SMI is a *function* of the probability distribution $p_1, \ldots, p_n$ associated with X.

The two experiments (or rv's) X and Y are said to be *independent* when the following equality holds for each x and y:

$$p(x, y) = p(x)p(y). \tag{3.4}$$

The first theorem about $H(X, Y)$ is as follows:

If X and Y are independent, then

$$H(X, Y) = H(X) + H(Y), \tag{3.5}$$

i.e. the SMI defined on the joint probabilities $p(x, y)$ is the sum of the SMI defined on the probability distribution $p(x)$ and the SMI defined on the probability distribution $p(y)$. The equality (3.5) follows straightforwardly from the definitions of Eqs. (3.1) and (3.4)

$$
\begin{aligned}
H(X,Y) &= -\sum_{x,y} p(x,y)\log p(x,y) \\
&= -\sum_{x,y} p(x)p(y)\log p(x)p(y) \\
&= -\sum_{x,y} p(x)p(y)\log p(x) - \sum_{x,y} p(x)p(y)\log p(y) \\
&= H(X) + H(Y).
\end{aligned}
\tag{3.6}
$$

Note that in the last step of Eq. (3.6) we have used the two conditions $\sum_x p(x) = 1$ and $\sum_y p(y) = 1$.

Next, we define the *conditional* SMI of X *given* that the value of Y is y, by

$$
H(X|Y = y) = -\sum_{x} p(x|y)\log p(x|y). \tag{3.7}
$$

Note that in this equation we have used the conditional probability of the event $X = x$, given the event $Y = y$. The summation in the equation is over all values of x.

The conditional SMI of X *given* Y is defined as the *average* of $H(X|Y = y)$ over all possible values of y, i.e.

$$
\begin{aligned}
H(X|Y) &= \sum_{x} p(y)H(X|Y = y) \\
&= -\sum_{y} p(y)\sum_{x} p(x|y)\log p(x|y).
\end{aligned}
\tag{3.8}
$$

Using the definition of the conditional probability

$$
p(x|y) = \frac{p(x,y)}{p(y)}, \tag{3.9}
$$

we rewrite Eq. (3.8) as

$$
H(X|Y) = -\sum_{x,y} p(x,y)\log\left[\frac{p(x,y)}{p(y)}\right] = H(X,Y) - H(Y). \tag{3.10}
$$

Rearranging Eq. (3.10), we get

$$H(X, Y) = H(X|Y) + H(Y). \tag{3.11}$$

Similarly, we can show, by exchanging the roles of X and Y in the derivation of (3.10) or (3.11), that

$$H(X, Y) = H(Y|X) + H(X). \tag{3.12}$$

We interpret Eqs. (3.5), (3.11), and (3.12) as follows. When X and Y are independent, the uncertainty about the outcome of both the experiments X and Y is simply the sum of the uncertainties regarding X and Y. This is consistent with the intuitive notion of a measure of uncertainty. It is also consistent with the intuitive interpretation of H as a measure of the information, i.e. the information about both X and Y is simply the sum of the information about X and the information about Y.

When the two experiments are *not independent*, we intuitively feel that knowing something about X might tell us something about Y, and vice versa. We can interpret Eq. (3.11) in terms of uncertainty as follows. The uncertainty about X and Y is the uncertainty about Y *plus* the uncertainty about X given that we know Y. In terms of a measure of information, we interpret Eq. (3.11) as the amount of information contained in the experiment Y plus the additional information on X given that we know Y. A similar interpretation holds for Eq. (3.12).

Pause to reflect on the restricted meaning of "uncertainty" and "information." These are always defined with respect to some probability distribution.

We next prove an important inequality:

For any two experiments or two rv's X and Y, we have

$$H(X, Y) \leq H(X) + H(Y), \tag{3.13}$$

where the inequality holds if and only if X and Y are independent. To prove the inequality, we start with the definitions of $H(X)$ and

$H(Y)$ in the form

$$H(X) = -\sum_{x} p(x) \log p(x) = -\sum_{x}\sum_{y} p(x,y) \log p(x), \qquad (3.14)$$

$$H(Y) = -\sum_{y} p(y) \log p(y) = -\sum_{x}\sum_{y} p(x,y) \log p(y). \qquad (3.15)$$

Note that $p(x) = p_X(x) = p(X = x)$ and $p(y) = p_Y(y) = p(Y = y)$. The summations in Eqs. (3.14) and (3.15) are over all possible values of x and y. Also, we used Eqs. (3.2) and (3.3) to rewrite the definitions of $H(X)$ and $H(Y)$ as sums over all possible values of both x and y. From Eqs. (3.14) and (3.15) we write

$$\begin{aligned} H(X) + H(Y) &= -\sum_{x}\sum_{y} p(x,y)[\log p(x) + \log p(y)] \\ &= -\sum_{x}\sum_{y} p(x,y) \log p(x)p(y). \end{aligned} \qquad (3.16)$$

We now apply the inequality (E.18) from Appx. E to get the inequality

$$H(X) + H(Y) - H(X,Y) = \sum_{x}\sum_{y} p(x,y) \log \frac{p(x,y)}{p(x)p(y)} \geq 0. \qquad (3.17)$$

Note that the double sum in Eq. (3.17) is over all possible values of both x and y. Clearly, when X and Y are independent, $p(x,y) = p(x)p(y)$, for each x and y, and we have the equality in (3.17); see Eq. (3.5). On the other, since $p(x,y)$ are probabilities, i.e. $p(x,y) \geq 0$, an equality in (3.17) implies that X and Y are independent.

Thus, we see that the difference on the left hand side of Eq. (3.17) is a measure of the extent of dependence between X and Y. We define the *correlation* between the events $X = x$ and $Y = y$ by

$$g(x,y) = \frac{p(x,y)}{p(x)p(y)}. \qquad (3.18)$$

When $g(x, y) = 1$, we say that there is no correlation between the two events in the sense that

$$p(x, y) - p(x)p(y) = 0. \tag{3.19}$$

When $g(x, y) > 1$, we say that the correlation is *positive* in the sense that

$$p(x, y) - p(x)p(y) > 0. \tag{3.20}$$

And when $0 < g(x, y) < 1$, we say that the correlation is negative in the sense that

$$p(x, y) - p(x)p(y) < 0. \tag{3.21}$$

Another way of interpreting the correlation is in terms of the conditional probabilities. We write

$$g(x, y) = \frac{p(x, y)}{p(x)p(y)} = \frac{p(x|y)}{p(x)}. \tag{3.22}$$

No correlation means knowing the occurrence of y has no effect on the probability of occurrence of x, i.e.

$$p(x|y) = p(x). \tag{3.23}$$

Positive correlation means knowing the occurrence of y *increases* the probability that x occurs, i.e. $g(x, y) > 1$, or

$$p(x|y) > p(x). \tag{3.24}$$

Negative correlation means knowing the occurrence of y *decreases* the probability that x occurs, i.e. $g(x, y) < 1$, or $p(x|y) < p(x)$.

3.2 Example: The Urn Problem

We present here a simple example for which we can calculate all the probabilities and all the relevant SMI's. The reader is urged to follow all the details, pause after each result, and think about its meaning.

Fig. 3.1. An urn with two white and three black marbles.

In an urn, there are two white marbles and three black marbles, see Fig. 3.1. The first experiment, X_{I}, is the draw of the first marble from the urn. The drawn marble is not returned. The second experiment, X_{II}, is the draw of the second marble from the urn, after the first marble was drawn, and it is not returned.

We now calculate the SMI of the two experiments, the conditional information, and all the relevant correlations, which we will use in Sec. 3.3 to calculate the mutual information. We use the symbols W and B for white and black. The following probabilities can be calculated by inspection of Fig. 3.1. We assume that the probabilities of drawing a white or a black marble depend only on the numbers of white and black marbles in the urn:

$$p(W_{\mathrm{I}}) = \frac{2}{5}, \quad p(B_{\mathrm{I}}) = \frac{3}{5}. \tag{3.25}$$

The conditional probabilities are

$$p(W_{\mathrm{II}}|W_{\mathrm{I}}) = \frac{1}{4}, \quad p(B_{\mathrm{II}}|W_{\mathrm{I}}) = \frac{3}{4}, \tag{3.26}$$

$$p(W_{\mathrm{II}}|B_{\mathrm{I}}) = \frac{1}{2}, \quad p(B_{\mathrm{II}}|B_{\mathrm{I}}) = \frac{1}{2}. \tag{3.27}$$

The probability of obtaining W in the second draw is

$$p(W_{\rm II}) = p(W_{\rm II}|W_{\rm I})p(W_{\rm I}) + p(W_{\rm II}|B_{\rm I})p(B_{\rm I}) = \frac{1}{4}\times\frac{2}{5}+\frac{1}{2}\times\frac{3}{5} = \frac{2}{5}. \quad (3.28)$$

The probability of obtaining B in the second draw is

$$p(B_{\rm II}) = p(B_{\rm II}|B_{\rm I})p(B_{\rm I}) + p(B_{\rm II}|W_{\rm I})p(W_{\rm I}) = \frac{1}{2}\times\frac{3}{5}+\frac{3}{4}\times\frac{2}{5} = \frac{3}{5}. \quad (3.29)$$

The probabilities $p(W_{\rm II})$ and $p(B_{\rm II})$ are calculated by the *total probability theorem*. Note carefully the equality:

$$p(W_{\rm I}) = p(W_{\rm II}), p(B_{\rm I}) = p(B_{\rm II}). \quad (3.30)$$

Next, we calculate the "inverse" probabilities. Here, we use the definition of the conditional probability.

$$p(W_{\rm I}|W_{\rm II}) = \frac{p(W_{\rm II}|W_{\rm I})p(W_{\rm I})}{p(W_{\rm II})} = \frac{\frac{1}{4}\times\frac{2}{5}}{\frac{2}{5}} = \frac{1}{4}, \quad (3.31)$$

$$p(B_{\rm I}|W_{\rm II}) = \frac{p(W_{\rm II}|B_{\rm I})p(B_{\rm I})}{p(W_{\rm II})} = \frac{\frac{1}{2}\times\frac{3}{5}}{\frac{2}{5}} = \frac{3}{4}, \quad (3.32)$$

$$p(W_{\rm I}|B_{\rm II}) = \frac{p(B_{\rm II}|W_{\rm I})p(W_{\rm I})}{p(B_{\rm II})} = \frac{\frac{3}{4}\times\frac{2}{5}}{\frac{3}{5}} = \frac{1}{2}, \quad (3.33)$$

$$p(B_{\rm I}|B_{\rm II}) = \frac{p(B_{\rm II}|B_{\rm I})p(B_{\rm I})}{p(B_{\rm II})} = \frac{\frac{1}{2}\times\frac{3}{5}}{\frac{3}{5}} = \frac{1}{2}. \quad (3.34)$$

.

Note carefully that these are "inverse" probabilities in the sense that we reverse the *order* of the events. Compare these probabilities with the conditional probabilities we calculated above, i.e. Eqs. (3.26) and (3.27).

The correlations between the various events are

$$g(W_{\mathrm{I}}, W_{\mathrm{II}}) = \frac{p(W_{\mathrm{II}}|W_{\mathrm{I}})}{p(W_{\mathrm{II}})} = \frac{p(W_{\mathrm{I}}|W_{\mathrm{II}})}{p(W_{\mathrm{I}})} = \frac{\frac{1}{4}}{\frac{2}{5}} = \frac{5}{8}, \tag{3.35}$$

$$g(W_{\mathrm{I}}, B_{\mathrm{II}}) = \frac{p(B_{\mathrm{II}}|W_{\mathrm{I}})}{p(B_{\mathrm{II}})} = \frac{p(W_{\mathrm{I}}|B_{\mathrm{II}})}{p(W_{\mathrm{I}})} = \frac{\frac{1}{2}}{\frac{2}{5}} = \frac{5}{4}, \tag{3.36}$$

$$g(B_{\mathrm{I}}, B_{\mathrm{II}}) = \frac{p(B_{\mathrm{II}}|B_{\mathrm{I}})}{p(B_{\mathrm{II}})} = \frac{p(B_{\mathrm{I}}|B_{\mathrm{II}})}{p(B_{\mathrm{I}})} = \frac{\frac{1}{2}}{\frac{3}{5}} = \frac{5}{6}, \tag{3.37}$$

$$g(B_{\mathrm{I}}, W_{\mathrm{II}}) = \frac{p(W_{\mathrm{II}}|B_{\mathrm{I}})}{p(W_{\mathrm{II}})} = \frac{p(B_{\mathrm{I}}|W_{\mathrm{II}})}{p(B_{\mathrm{I}})} = \frac{\frac{3}{4}}{\frac{3}{5}} = \frac{5}{4}. \tag{3.38}$$

Note that two of the correlations are positive ($g > 1$) and two negative ($g < 1$). If you know that W_{I} occurred, the probability of occurrence of W_{II} *decreases*, while that of B_{II} *increases*. Similarly, knowing that B_{I} occurred *decreases* the probability of B_{II} but *increases* the probability of W_{II}.

Pause and rationalize these results.

The SMI's of the two experiments are (all results are in bits)

$$H(X_{\mathrm{I}}) = -p(W_{\mathrm{I}})\log p(W_{\mathrm{I}}) - p(B_{\mathrm{I}})\log p(B_{\mathrm{I}}) = 0.971, \tag{3.39}$$

$$H(X_{\mathrm{II}}) = -p(W_{\mathrm{II}})\log p(W_{\mathrm{II}}) - p(B_{\mathrm{II}})\log p(B_{\mathrm{II}}) = 0.971. \tag{3.40}$$

Note that the SMI for both X_{I} and X_{II} is 0.971 bits. The conditional SMI's are

$$H(X_{\mathrm{II}}|W_{\mathrm{I}}) = -p(W_{\mathrm{II}}|W_{\mathrm{I}})\log p(W_{\mathrm{II}}|W_{\mathrm{I}}) \\ -p(B_{\mathrm{II}}|W_{\mathrm{I}})\log p(B_{\mathrm{II}}|W_{\mathrm{I}}) = 0.811, \tag{3.41}$$

$$H(X_{\mathrm{II}}|B_{\mathrm{I}}) = -p(W_{\mathrm{II}}|B_{\mathrm{I}})\log p(W_{\mathrm{II}}|B_{\mathrm{I}}), \\ -p(B_{\mathrm{II}}|B_{\mathrm{I}})\log p(B_{\mathrm{II}}|B_{\mathrm{I}}) = 1, \tag{3.42}$$

$$H(X_{\mathrm{II}}|X_{\mathrm{I}}) = p(B_{\mathrm{I}})H(X_{\mathrm{II}}|B_{\mathrm{I}}) + p(W_{\mathrm{I}})H(X_{\mathrm{II}}|W_{\mathrm{I}}) = 0.924, \tag{3.43}$$

$$H(X_{\mathrm{I}}|W_{\mathrm{II}}) = -p(W_{\mathrm{I}}|W_{\mathrm{II}})\log p(W_{\mathrm{I}}|W_{\mathrm{II}}) \\ -p(B_{\mathrm{I}}|W_{\mathrm{II}})\log p(B_{\mathrm{I}}|W_{\mathrm{II}}) = 0.811, \tag{3.44}$$

$$H(X_{\rm I}|B_{\rm II}) = -p(W_{\rm I}|B_{\rm II})\log p(W_{\rm I}|B_{\rm II}) \\ -p(B_{\rm I}|B_{\rm II})\log p(B_{\rm I}|B_{\rm II}) = 1, \tag{3.45}$$

$$H(X_{\rm I}|X_{\rm II}) = p(B_{\rm II})H(X_{\rm I}|B_{\rm II}) + p(W_{\rm II})H(X_{\rm I}|W_{\rm II}) = 0.924. \tag{3.46}$$

Note that if we know that $B_{\rm I}$ occurred, the conditional information about $X_{\rm II}$ is 1 bit. The same is true when we know that $B_{\rm II}$ occurred, and we ask for the conditional information about $X_{\rm I}$. Look at Fig. 3.1 and explain why this is true. The joint information is

$$H(X_{\rm I}, X_{\rm II}) = H(X_{\rm I}) + H(X_{\rm II}|X_{\rm I}) = H(X_{\rm II}) + H(X_{\rm I}|X_{\rm II}) = 1.895. \tag{3.47}$$

Exercise 3.1

Calculate all the quantities of the example above for the general case of N_W white and N_B black marbles. Note and explain some of the symmetries in the results.

Exercise 3.2

Repeat the same calculations for the case $N_W = 1$ and $N_B = 2$. Explain intuitively the meaning of the results.

3.3 Mutual Information

In this section, we introduce a concept which is very useful in communication theory, and which is considered to be more fundamental than SMI.

We define the mutual information between X and Y by

$$I(X,Y) = H(X) + H(Y) - H(X,Y) = \sum_x \sum_y p(x,y)\log\left[\frac{p(x,y)}{p(x)p(y)}\right] \\ = \sum_x \sum_y p(x,y)\log g(x,y) \geq 0. \tag{3.48}$$

The inequality for $I(X, Y)$ in Eq. (3.48) follows from the inequality (3.13). This is an important inequality; it means that the amount of information in the joint experiments (X, Y) cannot exceed the sum of the amount of information associated with X and with Y, separately.

We recall that the events $X = x$ and $Y = y$ are said to be positively correlated when $g(x, y) > 1$. In terms of the conditional probability, we write

$$p(X = x|Y = y) = \frac{p(X = x; Y = y)}{p(Y = y)} = g(x, y)p(X = x). \qquad (3.49)$$

Thus, positive correlation means that the conditional probability of x given y is *larger* than the probability of x.

Similarly, when $g(x, y) < 1$, we say that we have *negative* correlation, which means that the conditional probability $p(X = x|Y = y)$ is *smaller* than $p(X = x)$.

Note that $g(x, y)$ is defined *symmetrically* with respect to x and y. This means that if the occurrence of the event $X = x$ enhances the probability of the event $Y = y$, then the occurrence of the event $Y = y$ also enhances the probability of the event $X = x$. From Eq. (3.48), we see that the average of the logarithm of the correlation is always *positive*, although each term in the sum can be either positive or negative.

When X and Y are *dependent*, the correlation $g(x, y)$ can be either larger or smaller than one for any pair of events x and y. This means that $\log g(x, y)$ can be either positive or negative, for specific events x and y. However, the *average* of $\log g(x, y)$ as defined in Eq. (3.48) must always be positive.

The inequality (3.48) may be rewritten, using Eqs. (3.11) and (3.12), as

$$I(X; Y) = H(X) - H(X|Y) = H(Y) - H(Y|X) \geq 0. \qquad (3.50)$$

From Eqs. (3.48) and (3.50), we get the important result

$$H(X, Y) = I(X; Y) + H(Y|X) + H(X|Y). \qquad (3.51)$$

The inequality (3.50) suggests an important and informative interpretation of the quantity $I(X;Y)$. $H(X)$ is the uncertainty in the experiment X. $H(X|Y)$ is the uncertainty in X given that we know the result of the experiment Y. The inequality (3.50) means that knowing Y always *decreases* the uncertainty about X. Similarly, knowing X always decreases the uncertainty about Y(remember "uncertainty" with respect to want). In terms of information, we can also say that $I(X;Y)$ measures the amount of information *contained* in an experiment X with respect to an experiment Y, and vice versa, or the information *conveyed* by Y about X, and vice versa.

Two extreme cases are:

(1) When X and Y are independent, it follows from the definition (3.48) that $I(X;Y) = 0$. This means that no information is provided by X, knowing Y, and vice versa (or the message received is totally unrelated to the message sent).

(2) When $X = Y$,

$$H(X,Y) = H(X) = H(Y) \tag{3.52}$$

or, equivalently,

$$I(X;X) = H(X). \tag{3.53}$$

In this case, knowing X provides the maximum information about X. For this reason, $I(X;X)$ is sometimes referred to as self-information (note that this term is used for another quantity discussed in Sec.1.2).

Figure 3.2 shows the relationship between $H(X,Y), H(Y), H(Y)$, $H(X|Y)H(Y|X)$ and $I(X;Y)$.

Qualitatively, the inequalities (3.13) and (3.48) are intuitively clear. If we have two experiments, then knowing the result of one experiment can only add some information on the second. If the two experiments are independent, then knowing the outcome of one does not tell us anything about the other.

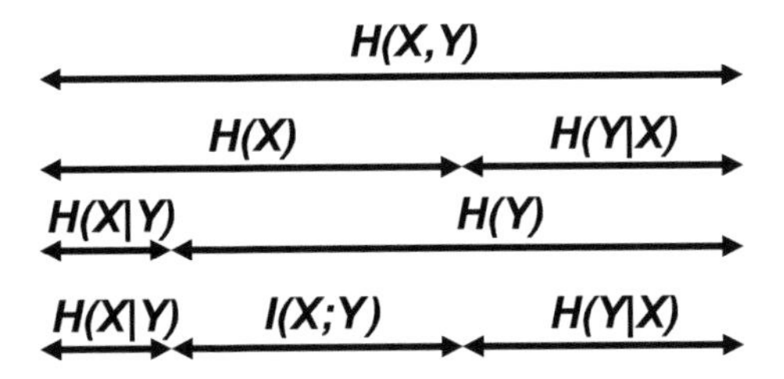

Fig. 3.2. The relationships between the quantities $H(X)$, $H(Y)$, $H(X|Y)$, $H(Y|X)$, $H(X,Y)$ and $I(X;Y)$.

We now return to the urn problem. The mutual information for the urn problem discussed in Sec. 3.2 is

$$I(X_{\mathrm{I}};X_{\mathrm{II}}) = H(X_{\mathrm{I}}) + H(X_{\mathrm{II}}) - H(X_{\mathrm{I}},X_{\mathrm{II}})$$
$$= 2\times 0.971 - 1.897 = 0.045 \text{ bits.} \quad (3.54)$$

In Eqs. (3.35)–(3.38), we found two positive and two negative correlations. The average of the logarithm of all these correlations, $I(X;Y)$, as defined in Eq. (3.48) must be *positive*. In general, one correlation can be larger than 1, and the second, smaller than 1. However, in forming the average in $I(X_{\mathrm{I}};X_{\mathrm{II}})$, the positive correlations get the larger weight. Therefore, the net effect is that $I(X_{\mathrm{I}};X_{\mathrm{II}})$ is always positive.

The inequality (3.48) is a special case of a more general inequality. For any two distributions defined on the same alphabet $X; p(x)$ and $q(x)$, the following inequality holds:

$$D(p\|q) = \sum_{x\in X} p(x)\log\left[\frac{p(x)}{q(x)}\right] \geq 0. \quad (3.55)$$

p and q denote the two distributions, and the sum is over all possible outcomes $x \in X$.

The quantity $D(p\|q)$ is referred to as the Kullback–Leibler information, or the *distance* between the two distributions p and q (and sometimes also as the "relative entropy"). Clearly, the mutual information is a special case of the inequality (3.55), i.e. $I(X;Y)$ is the *distance* between the two distributions $p(x, y)$ and $p(x)p(y)$. The quantity $D(p\|q)$ is very useful in applications of IT. We will not need it in this book.

Another important quantity in communication theory is the channel capacity. A channel is defined as a pair of rv's, X and Y, where X is the *input* for which the values x belong to one alphabet X_x, and Y the output for which the values y belong to a different alphabet X_Y. The channel is characterized by the conditional probabilities $P(Y = y|X = x)$, i.e. the probability of obtaining the letter $y \in X_Y$ as an output, given that the letter $x \in X_x$ was sent in the input.

Schematically, we describe a channel by the transformation

$$X \rightarrow P(Y = y|X = x) \rightarrow Y.$$

The pair of rv's X and Y defines the mutual information $I(X; Y)$. The channel capacity is defined by

$$C = \max_{p(X=x)} I(X; Y).$$

The maximum of $I(X; Y)$ is over all possible input distributions.

The capacity does not introduce any new concept in IT. However, this quantity is important in coding theory and the rate of transmission of information.

3.4 A System of Two Spins

In this section, we will clarify the meaning of all the SMI's we have defined above by means of a physical example. We will describe a physical system for which we can calculate all the relevant probabilities. With these probabilities we calculate all the SMI's defined in the previous sections of this chapter. We will also examine the physical meaning of all the results.

In all the examples, we consider two particles referred to as *spins*, placed at some fixed distance from each other. The spins can be electric or magnetic dipoles. We will not be interested here in the physical

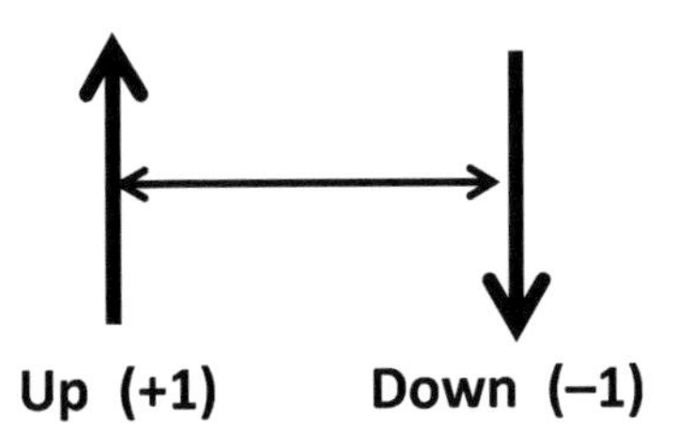

Fig. 3.3. Two states of a spin: "Up" and "Down."

nature of the particles. We provide only the method of calculating the probabilities of this system. Each spin can be in one of two states, referred to as "up" and "down," and assigned the numerical values of 1 and -1, respectively; see Fig. 3.3.

Thus, we have two experiments (or two rv's) X and Y; each can have one of two values, either $(+1)$, or (-1). In all we have four possible states of the entire system: $(1,1),(1,-1),(-1,1)$, and $(-1,-1)$.

The probabilities of these events are derived from the *potential energy* function defined by

$$U(x,y) = -J(x \times y) - F_1 x - F_2 y. \tag{3.56}$$

All the probabilities are calculated from the expression

$$p(x,y) = \frac{\exp[-\beta U(x,y)]}{z_2}, \tag{3.57}$$

where the normalization constant Z_2 is defined by

$$Z_2 = \sum_x \sum_y \exp[-\beta U(x,y)]. \tag{3.58}$$

Here, $\beta = (k_B T)^{-1}$, where k_B is the Boltzmann constant and T is the absolute temperature. We will use $k_B = 1$ and express T in absolute numbers.

Clearly, the recipe for calculating the probabilities from Eq. (3.57) is motivated by the Boltzmann distribution. There are four states of the system of the two spins (Fig. 3.4). Each state (x,y) is characterized by the energy $U(x,y)$, which determines its probability by Eq. (3.57).

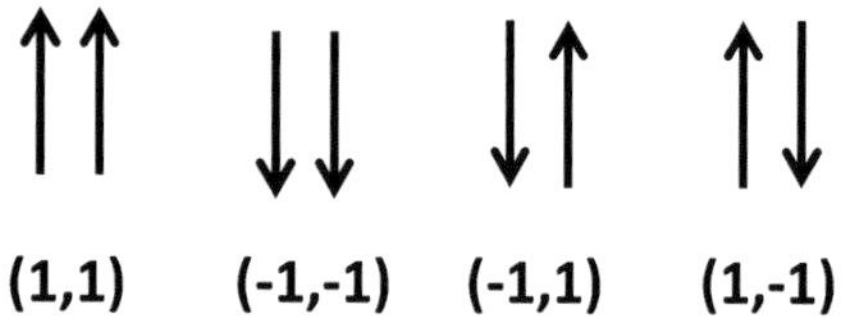

↑↑ ↓↓ ↓↑ ↑↓

(1,1) (-1,-1) (-1,1) (1,-1)

Fig. 3.4. The four states of the two spins.

The quantity J measures the *strength* of the *interaction* between the two spin ($J > 0$ corresponds to the case of ferromagnetism, and $J < 0$ to the case of antiferromagnetism). In all the examples used below, we choose either $J = 1$ or $J = -1$. F_1 and F_2 are referred to as the *external fields* operating on the two spins, respectively. In the first case (Sec. 3.4.1), we choose $F_1 = F_2 = 0$, and examine only the effect of the spin–spin interaction on the probabilities and the various SMI's of this system. Clearly, in this case, the two states $(1, 1)$ and $(-1, -1)$ will have the same interaction energy, i.e.

$$U(1,1) = U(-1,-1) = -J, \tag{3.59}$$

and also we have

$$U(1,-1) = U(-1,1) = J. \tag{3.60}$$

Thus, in this case, there are only two different probabilities for the system.

In the second case (Sec. 3.4.2), we introduce an external field, $F = F_1 = F_2$. This means that the same field F affects both of the spins. In this case, the equality (3.59) does not hold. Instead, we have three different joint probabilities in the system, corresponding to the energies

$$\begin{aligned} U(1,1) &= -J - F - F, \\ U(-1,-1) &= -J + F + F, \\ U(1,-1) = U(-1,1) &= J + F - F = J. \end{aligned} \tag{3.61}$$

In the third case (Sec. 3.4.3), we apply *different* fields to the two spins. Clearly, in this case, all the states shown in Fig. 3.4 will have different probabilities. In particular, instead of the last equality in (3.61), we will have

$$\begin{aligned} U(1,-1) &= J + F_1 - F_2, \\ U(-1,1) &= J - F_1 + F_2. \end{aligned} \tag{3.62}$$

Thus, in this case, we have four different joint probabilities.

3.4.1 *Two Interacting Spins without an External Field*

We start with the case of no spin–spin interaction.

If $J = 0$, the two experiments X and Y are independent. In this case, the probabilities are

$$p_X(1) = p_X(-1) = p_Y(1) = p_Y(-1) = \frac{1}{2}, \tag{3.63}$$

$$p_{X,Y}(1,1) = p_{X,Y}(-1,-1) = p_{X,Y}(1,-1) = p_{X,Y}(-1,1) = \frac{1}{4}. \tag{3.64}$$

The SMI of a single spin is

$$H(X) = H(Y) = -\sum_{x=\pm 1} p(x) \log p(x) = 1 \text{ bit}, \tag{3.65}$$

$$H(X,Y) = -\sum_{x=\pm 1} \sum_{x=\pm 1} p(x,y) \log p(x,y) = 2 \,\text{bits}. \tag{3.66}$$

As expected, the SMI for one spin is one bit, and for the two spins we have four states of equal probabilities, and hence the SMI is 2 bits. The mutual information in this case is

$$I(X;Y) = 0. \tag{3.67}$$

Pause and make sure you understand these results before proceeding to the next case.

Turning on the interaction

For the case $J = 1$ and $T = 1$, we have the following four pair four probabilities:

Y \ X	1	−1	P_X
1	0.44	0.06	0.5
−1	0.06	0.44	0.5
P_Y	0.5	0.5	

(3.68)

Joint probabilities and marginal probabilities for the case $J = 1$ and $T = 1$.

As expected, the states $(1, 1)$ and $(-1, -1)$ have the same probabilities, and these are much larger than the probabilities of the states $(1, -1)$ and $(-1, 1)$. The reason is clear when $J = 1$ the interaction between the two spins favors the *parallel* orientations over the antiparallel orientations of the two spins. Note also that the marginal probabilities of the two states are equal for each of the spins:

$$p_X(1) = p_Y(1) = p_X(-1) = p_Y(-1) = 0.5. \tag{3.69}$$

For the case $J = -1$ and $T = 1$, we have the following four joint probabilities:

Y \ X	1	−1	P_X
1	0.06	0.44	0.5
−1	0.44	0.06	0.5
P_Y	0.5	0.5	

(3.70)

Joint probabilities and marginal probabilities for the case $J = -1$ and $T = 1$.

In this case, the interaction between the spins favors the *antiparallel* orientations, i.e. $(1, -1)$ and $(-1, 1)$. The singlet probabilities are as in the previous case, Eq. (3.69).

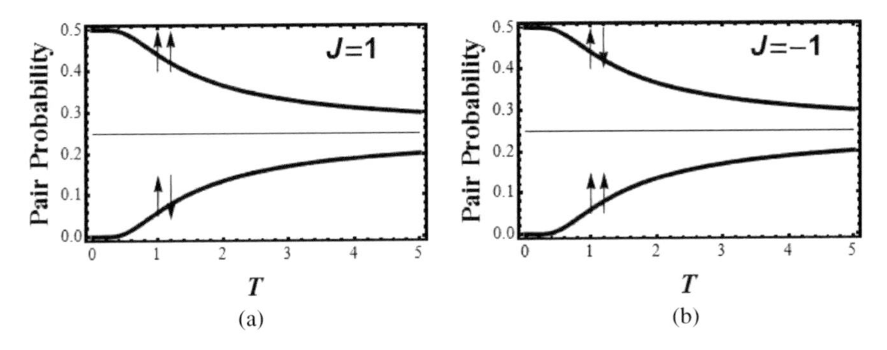

Fig. 3.5. The pair distributions as a function of T for (a) $J = 1$ and (b) $J = -1$.

Figure 3.5 shows how the pair probabilities change with temperature for the parallel orientation [both $(1,1)$ and $(-1,-1)$] and the antiparallel orientation [both $(1,-1)$ and $(-1,1)$].

For $J = 1$, Fig. 3.5(a) shows that the parallel configuration has a larger probability than the antiparallel for all temperatures. At $T = 0$, the probability of each of the parallel configurations is

$$P(1,1) = P(-1,-1) = 0.5. \tag{3.71}$$

On the other hand, for the antiparallel configurations $(1,-1)$ and $(-1,1)$,

$$P(1,-1) = P(-1,1) = 0. \tag{3.72}$$

As the temperature increases, the two curves approach each other, and at $T \to \infty$, all the pair probabilities approach the limit of 1/4, i.e. equal probabilities for the four states.

We find similar behavior for the case $J = -1$ in Fig. 3.5(b), except that the roles of the parallel and antiparallel configurations are reversed, i.e. here the antiparallel configuration is always the more probable one.

Note that the equality (3.69) for the marginal probabilities is maintained for any temperature.

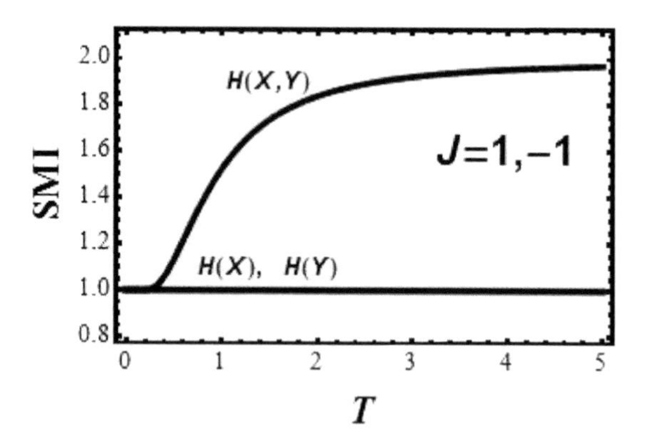

Fig. 3.6. The three SMIs for the system of two spins in the absence of a field.

Figure 3.6 shows the SMI for X, Y, and for (X, Y). As expected, both $H(X)$ and $H(Y)$ have a constant value of 1 bit, independent of T and J. On the other hand, the value of $H(X, Y)$ starts at 1 bit at $T = 0$. This is a result of having only two states with probabilities 1/2; see Eqs. (3.71) and (3.72). As the temperature increases, the SMI for the pair (X, Y) also increases. This is a result of the fact that at nonzero temperatures there are four configurations having nonzero probabilities, and as the temperature increases all these probabilities tend to be equal (see Fig. 3.5). Hence, the limiting value of $H(X, Y)$ at large T is 2 bits:

$$H(X, Y) \underset{T\to\infty}{\longrightarrow} -4\left(-\frac{1}{4}\log\frac{1}{4}\right) = 2 \text{ bits.} \tag{3.73}$$

Exercise 3.3

Calculate the conditional probability $P(X = 1|Y = 1)$ as a function of T for different values of the interaction. When we turn on the interaction $J > 0$ at $T = 0$, there is positive correlation, i.e. if one spin is "up," the probability of the second one to be "up" is 1. However, for any other temperatures the correlation is still positive, i.e. $P(X = 1|Y = 1) > P(X = 1) = 0.5$, but the value of the conditional probability decreases with T. For any given temperature, the larger

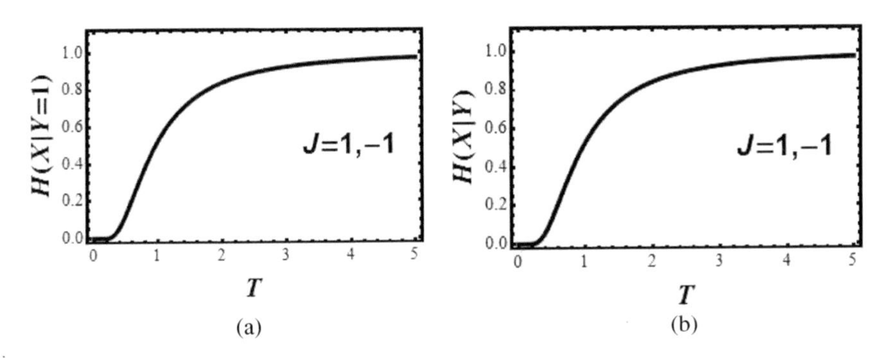

Fig. 3.7. The conditional SMI: (a) Given a specific outcome, and (b) average of the specific conditional SMIs.

the (positive) value of J, the larger the conditional probability, and for $J \to \infty$ the conditional probability will tend to 1.

Exercise 3.4

Calculate the values of $P(X = 1|Y = -1)$ as a function of T for different values of J. Note that in this case all the correlations are negative (for $J > 0$). The reason is that if one spin is "up," then the conditional probability of the second spin to be "up" is *less* than 0.5.

Next, we show in Fig. 3.7(a) the conditional information of either $H(X|Y = 1)$ or $H(X|Y = -1)$ as a function of T. Clearly, in this case, knowing the outcome of either $Y = 1$ or $Y = -1$ has the same effect on the uncertainty of X. Furthermore, the probability of finding either $Y = 1$ or $Y = -1$ is the same (0.5) for any T [Eq. (3.69)]. Therefore, when we take the average in Eq. (3.8) we obtain $H(X|Y)$, which has the same value of either $H(X|Y = 1)$ or $H(X|Y = -1)$. This is true for both $J = 1$ and $J = -1$. As one can see by comparing Fig. 3.7(b) with Fig. 3.6, the shape of the curve of $H(X|Y)$ is the same as that of $H(X, Y)$ except that $H(X, Y)$ is larger by exactly 1 bit at any T. This is a result of the fact that in this example $H(X)$ and $H(Y)$ have a constant value of 1 bit, and the difference between $H(X, Y)$ and either $H(X|Y)$ or $H(Y|X)$ is 1 bit; see Eqs. (3.11) and (3.12).

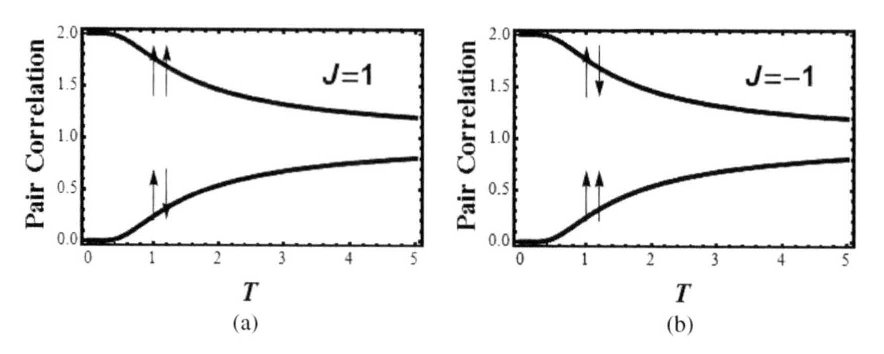

Fig. 3.8. Pair correlations: (a) For the case $J = 1$, and (b) for the case $J = -1$.

Next, we show the pair correlation defined in Eq. (3.18) as a function of T, for the case of $J = 1$ in Fig. 3.8(a) and for the case $J = -1$ in Fig. 3.8(b). The two figures are identical except for the change in the roles of parallel and antiparallel configurations. For $J = 1$, the correlation between the two parallel spins is positive. This means that once we know that the state of one spin is "up" (or "down"), the conditional probability of the second spin to be "up" (or "down") is larger than the probability of "up" (or "down"); see Eq. (3.20). This effect is a result of the interaction between the spins which favors the parallel [either $(1, 1)$ or $(-1, -1)$] configurations. On the other hand, given that one spin is "up" (or "down"), the conditional probability of finding the second one in the state "down" (or "up") is smaller than the probability of "up" (or "down"). Note also that in each of the Figs. 3.8(a) and 3.8(b), the positive correlation starts at 2 while the negative correlation starts at 0 when $T = 0$. Both correlations tend to 1 at $T \to \infty$, which means that at high temperatures there is no correlation between the two spins.

As we have shown in Eq. (3.48), although $\log g(x, y)$ may be either positive or negative for the specific events $X = x$ and $Y = y$, the *average* of $\log g(x, y)$ must be *positive*. Figure 3.9 shows the mutual information between the two spins as a function of the temperature. The curve is the same for $J = 1$ and $J = -1$. The value of $I^{(2)}(X; Y)$ starts at 1 bit at $T = 0$. This means that if we know the result of one spin, we know with certainty the result of the second spin. On the other hand, for $T \to \infty$ the mutual information is zero. This means that no matter what the interaction between the spins is, at very high temperatures there is no correlation between the states of the two spins. Knowing the state of one spin does not provide any information on the state of the other spin. Figure 3.10 shows $I^{(2)}(X;Y)$ as a function of T for different values of the parameter J. For $J = 0$, there is no interaction, and hence $I^{(2)} = 0$, i.e. there is no mutual information. When the interaction is "turned on," $J \neq 0$, we see that the larger the J, the larger the mutual information. When $J = 10$,

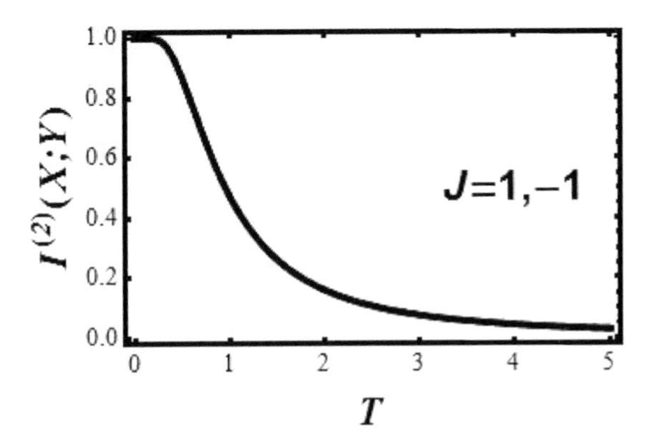

Fig. 3.9. The mutual information for the two spins in the absence of the field.

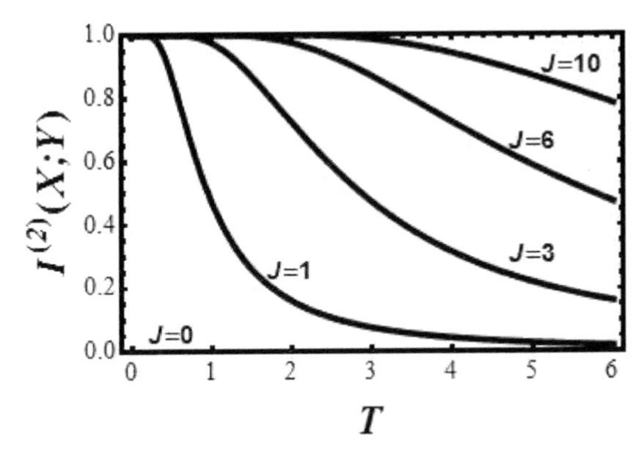

Fig. 3.10. The mutual information as a function of T for different values of J.

the mutual information is almost 1 (at low temperature). Knowing X gives us much information on Y, and vice versa. In Figs. 3.9 and 3.10, we have used the notation $I^{(2)}$ for the mutual information $I(X;Y)$ to emphasize that we are studying the mutual information between two spins in a system consisting of only two spins. In Chap. 4, we will study the mutual information between two spins in a system of three or more spins. In these cases, we will use the notation $I^{(3)}, I^{(4)}$, etc.

3.4.2 *Two Interacting Spins with an External Field F*

In this example, we add an external field F operating on the spin; see Fig. 3.11. The potential energy function in this case is

$$U(x,y) = -J(x \times y) - F \times x - F \times y. \tag{3.74}$$

Thus, the interaction between the two spins is as in the previous example, but in addition the external field F operates on each of the spins. By this choice we remove one degeneracy that we had in the previous example. If we choose $F > 0$, then the field will favor the configuration "up," and therefore the configurations $(1,1)$ and $(-1,-1)$ will no longer have the same probabilities as in the example in Sec. 3.4.1. Note, however, that the configurations $(1,-1)$ and $(-1,1)$ still have the same probability even in the presence of the field. We will remove

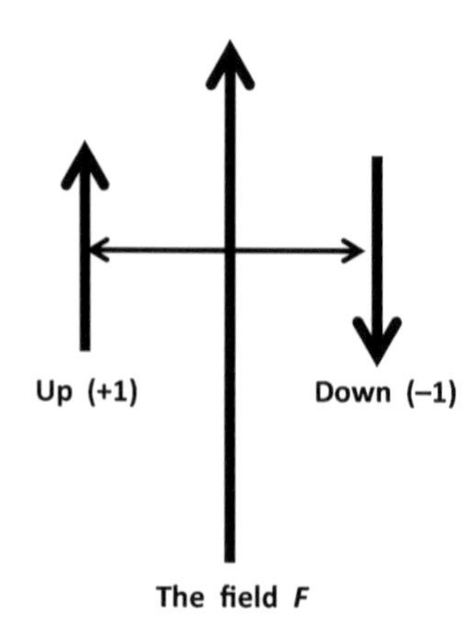

Fig. 3.11. The two spins in the presence of the field, F.

this last degeneracy in the next example. We will not study the case $F < 0$ separately, because it does not add anything new.

Once we have the modified potential function, Eq. (3.74), we can calculate all the relevant quantities using Eq. (3.73) instead of Eq. (3.56), but with $F_1 = F_2 = F$.

We start with the pair probabilities. For $T = 1, F = 0.1$, and $J = 1$ we have

Y \ X	1	−1	P_X
1	0.529	0.059	0.588
−1	0.059	0.353	0.412
P_Y	0.588	0.412	

(3.75)

Joint probabilities and marginal probabilities for the case $J = 1$ and $T = 1$ and $F = 0.1$.

Note that the field increases the probability of $(1, 1)$ and decreases the probability of $(-1, -1)$. Compare these values with the values in Sec. 3.4.1: Eqs. (3.68) and (3.70). Note also that the probabilities of $(1, -1)$ and $(-1, 1)$ are equal, as expected (explain why).

For $J = -1$ and $T = 1, F = 0.1$, the interaction energy favors the antiparallel configurations, i.e. $(-1, 1)$ and $(1, -1)$, and hence the probabilities in this case are

Y \ X	1	−1	P_X
1	0.073	0.439	0.512
−1	0.439	0.049	0.488
P_Y	0.512	0.488	

(3.76)

Joint probabilities and marginal probabilities for the case $J = 1$ and $T = 1$ and $F = 0.1$.

Note that in this case the external field still favors the (1,1) over the $(-1, -1)$ configuration. However, because of the interaction energy $J = -1$, the antiparallel configurations have the larger probability.

The singlet probabilities are also affected by the external field. Unlike the previous case [see Eq. (3.69)], here we have, for $J = 1$, $F = 0.1, T = 1$,

$$p_X(1) = p_Y(1) = 0.588,$$
$$p_X(-1) = p_Y(-1) = 0.412. \tag{3.77}$$

Note that for this relatively small external field ($F = 0.1$) the "up" state is slightly more favored than the "down" state. For $J = -1$, we have

$$p_X(1) = p_Y(1) = 0.512.$$
$$p_X(-1) = p_Y(-1) = 0.488. \tag{3.78}$$

Again, the "up" state is slightly more favored than the "down" state of a single spin.

Once we "turn on" the field, we get a very rich repertoire of behaviors of the system. This is already revealed in the singlet and the pair probabilities. Figure 3.12 shows the singlet probabilities for the "up" and "down" states as a function of the temperature T for different values of F.

First, note in Fig. 3.12(a) that when $F = 0$, $P(1)$ is constant, $P(1) = 0.5$, and does not change with T. Once we turn on the external field, the probability of the "up" state steadily increases.

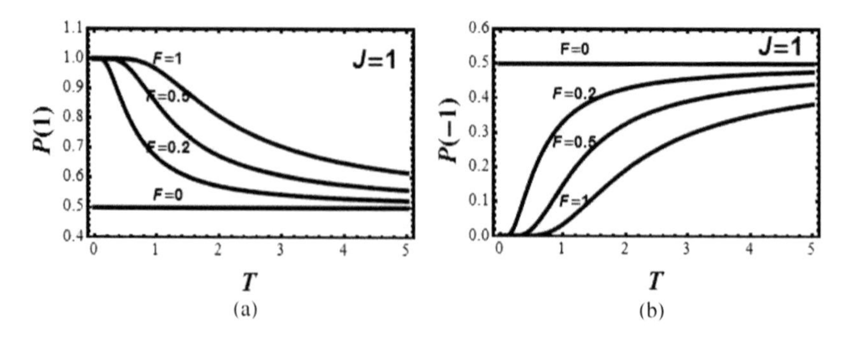

Fig. 3.12. The singlet probabilities as a function of T for: (a) The state (1), and (b) the state (−1).

The temperature dependence of $P(1)$ is also interesting. At $T = 0$, all the curves of $P(1)$ as a function of T start at $P(1) = 1$ (except for the one with $F = 0$). This is understandable. At very low temperature, even a small external field makes the orientation of the spin "up" most favored, i.e. with probability 1 [and the probability of "down" (i.e. -1) almost 0; see Fig. 3.12(b)]. As the temperature increases, the preference for the "up" state caused by the field diminishes, and at a high temperature all the curves in Figs. 3.12(a) and 3.12(b) tend to $P(1) = P(-1) = 0.5$. At this limit, the effect of the temperature overcomes the effect of the field, and the system is totally randomized. Similar results are shown in Fig. 3.12(b) for the probability of "down." Here, all the curves (except for the one with $F = 0$) start with probability 0 at $T = 0$, and then the probabilities climb up as the temperature increases, reaching $P(1) = 0.5$ at $T \to \infty$.

The case $J = -1$ is very different from the case $J = 1$. Figure 3.13 shows some data for $P(1)$ as a function of T, for small values of $F \leq 1$ in Fig. 3.13(a) and for larger values of $F \geq 1$ in Fig. 3.13(b).

Note that all the curves of $P(1)$, and for $F < 1$, start at $T = 0$ with $P(1) = 0.5$. A unique behavior occurs for $F = 1$, where we have, at $T = 0$, $P(1) = 2/3$. On the other hand, for $F > 1$, all curves start at $P(1) = 1$ at $T = 0$. Also note that for small values of F the curve of $P(1)$ as a function of T goes through a maximum. This can be explained by the competition between the effect of the interaction between the two spins and the effect of the interaction with the external field. Remember, if we had only *one* spin in an external field, the field would always enhance the probability of the "up" state. In our case, we have two spins. Let us focus on the left spin (the experiment X). The external field will favor the "up" state of X, but it also favors the "up" state of Y. However, because of the interaction ($J = -1$), the second spin Y will affect the spin X to orient itself opposite to X. Thus, the external field F tends to favor both X and Y to attain the "up" state. But the "up" state of X will enhance the probability of the "down" state of Y. These two competing effects produce the maximum in

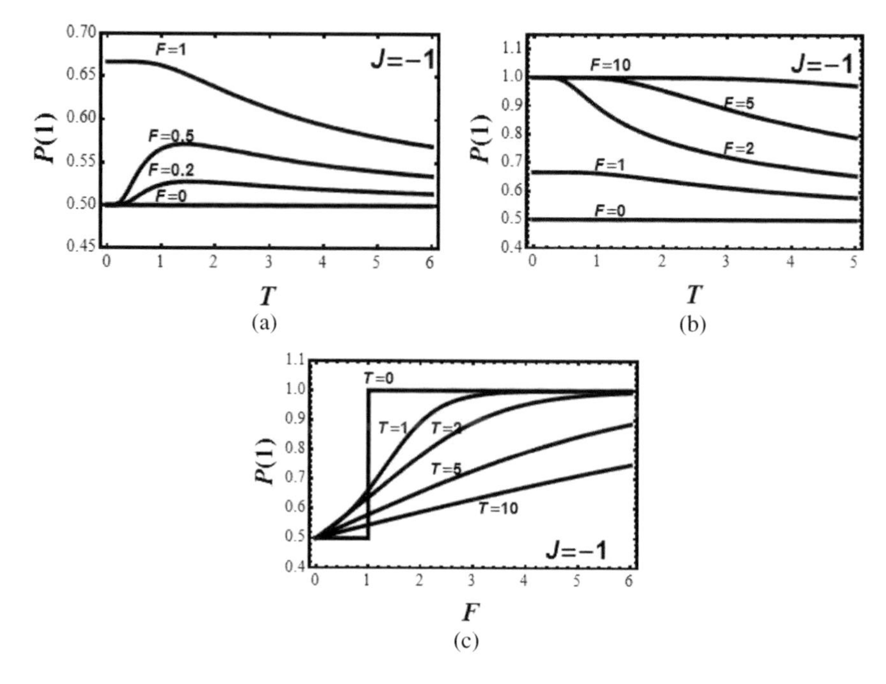

Fig. 3.13. The singlet probabilities $P(1)$ for negative J: (a) As a function of T for small F, (b) as a function of T for larger F, and (c) as a function of F for different T.

the curves in Fig. 3.13(a). For very large values of F, we start with $P(1) = 1$ at $T = 0$. Then the probability $P(1)$ decreases monotonically as T increases. A unique behavior is observed for $F = 1$, which starts with $P(1) = 2/3$ at $T = 0$, then decreases toward 0.5 at $T \to \infty$. In Fig. 3.13(c), we show the dependence of $P(1)$ on the strength of the field for various temperatures. We observe that at $F = 1$ and $T = 0$, there is a sharp transition from $P(1) = 0.5$ to $P(1) = 1$. At $T = 0$, the competition between the interaction (J) and the field (F) causes $P(1)$ to turn sharply from randomness $[P(1) = 0.5]$ at $F < 1$ to certainty $[P(1) = 1]$ at $F > 1$.

Figure 3.14 shows similar results for $P(-1)$. Here, for low values of F, we observe a minimum of $P(-1)$ as a function of T. For $F = 1$, we have a limit of $P(-1) = 1/3$ at $T = 0$, and for higher values

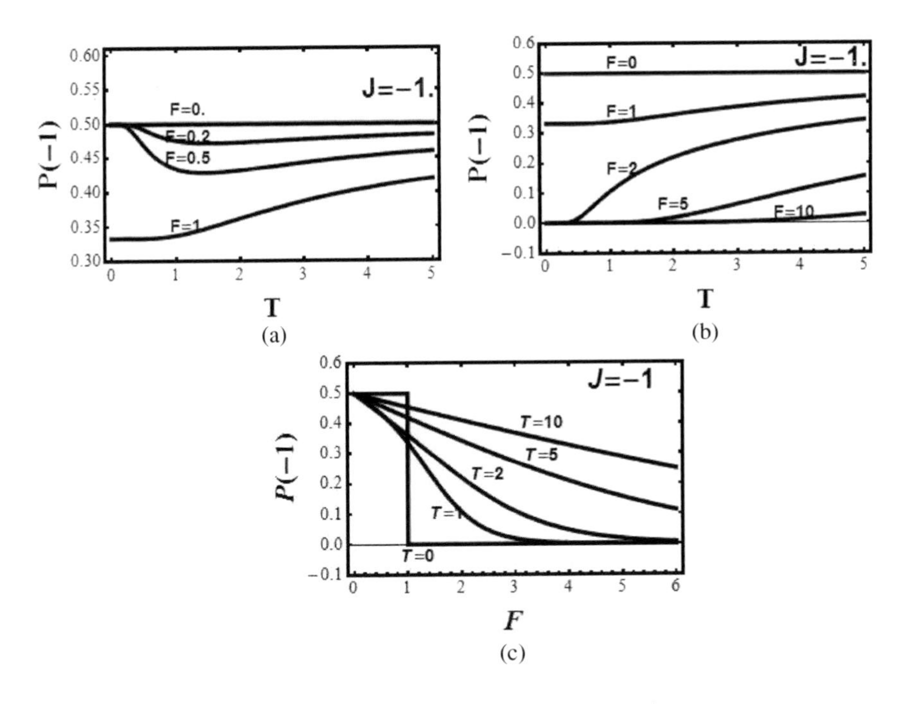

Fig. 3.14. Similar results as in Fig. 3.13 but for $P(-1)$.

of F the curves start at $P(-1) = 0$ and climb to $P(-1) = 0.5$ at very high temperatures. Figure 3.14(c) shows $P(-1)$ as a function of F for different temperatures. We see again a sharp transition from $P(-1) = 0.5$ to $P(-1) = 0$ at about $F = 1$.

The unique behavior of the case of $F = 1$ is a result of our choice of $J = -1$. In this case, the effect of the field on one spin cancels the effect of the interaction, i.e.

$$-J - F - F = 1 - 1 - 1 = -1. \tag{3.79}$$

If we choose, say, $J = -5$ and $F = 5$, we will get the same behavior of $P(1)$ and $P(-1)$ as a function of T.

For the pair probabilities, we find a rich variety of dependence on $T, J,$ and F.

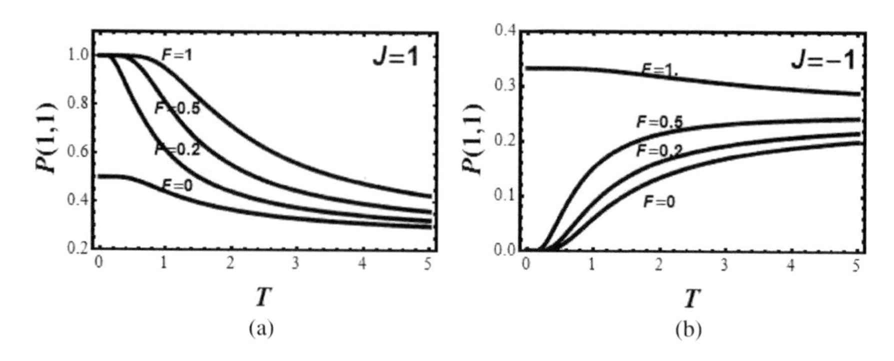

Fig. 3.15. Pair probabilities $P(1,1)$ of the two spins as a function of T for different values of F: (a) The case $J = 1$, and (b) the case of $J = -1$.

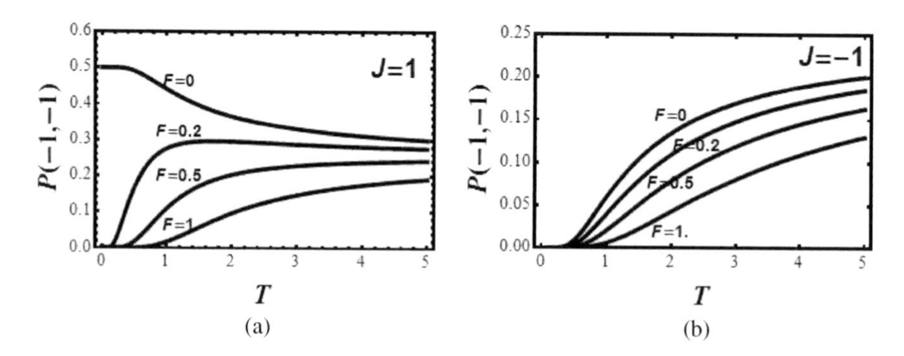

Fig. 3.16. Pair probabilities $P(-1,-1)$ of the two spins as a function of T for different values of F: (a) The case $J = 1$, and (b) the case of $J = -1$.

Figure 3.15 shows the behavior of $P(1,1)$ as a function of T for various values of F and for $J = 1, J = -1$. As we can see, the dependence of $P(1,1)$ on T varies considerably with the strength of the field and with the interaction (J).

Figure 3.16 shows the behavior of $P(-1,-1)$, and Fig. 3.17 the behavior of $P(1,-1) = P(-1,1)$ for $J = 1$ and $J = -1$. The reader is urged to try to examine these behaviors of the pair probabilities.

Figure 3.18(a) shows the SMI for a single spin as a function of T for values of F. In the case $J = 1$ and $F = 0$, $H(X) = 1$ independently of T. Once we turn on the field, the SMI starts at $H(X) = 0$ at $T = 0$

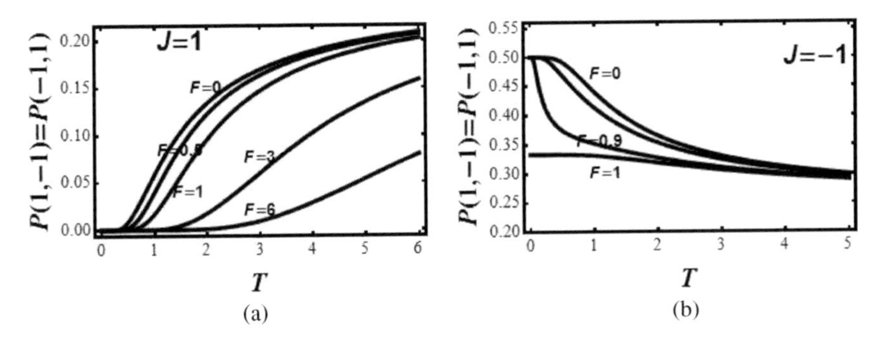

Fig. 3.17. Pair probabilities $P(-1,1)$ and $P(1,-1)$ of the two spins as a function of T for different values of F: (a) The case $J = 1$, and (b) the case of $J = -1$.

(even a small field at $T = 0$ causes certainty in the orientation of the spin), and as the temperature increases, the value of $H(X)$ increases (the uncertainty about the orientation increases with T for each F, but decreases with F at each T).

A different behavior is observed for $J = -1$; see Fig. 3.18(b). Here, we see that for small values of $F < 1$, the value of $H(X)$ goes through a minimum. The reason is similar to the one given above in connection with the singlet probabilities. In the case $J = -1$ and $F < 1$ [Fig. 3.18(b)], the field enhances the orientation of the "up" state for spin X, and for spin Y. However, there is a secondary effect of the orientation of Y on X which tends to orient X in the opposite direction to Y. These two effects compete with each other, such that at low T one effect dominates and at high temperature the second effect dominates. For larger values of $F > 1$ [Fig. 3.18(c)], $H(X)$ starts with $H(X) = 0$ at $T = 0$, and monotonically increases toward $H(X) = 1$ for higher temperatures. Here, because of the strong field $F > 1$, the effect of the field dominates the behavior of the spin. A unique behavior is observed for $F = 1$. At $T = 0$, the value of $H(X)$ tends to about 0.918. This is a result of the specific distribution of the two states, as we observed in Fig. 3.12(a) for $p(1)$ at $T = 0$.

Figure 3.19 shows the dependence of $H(X)$ on F for different temperatures. All the curves start at $F = 0$ with $H(X)$ equal to 1 bit. This

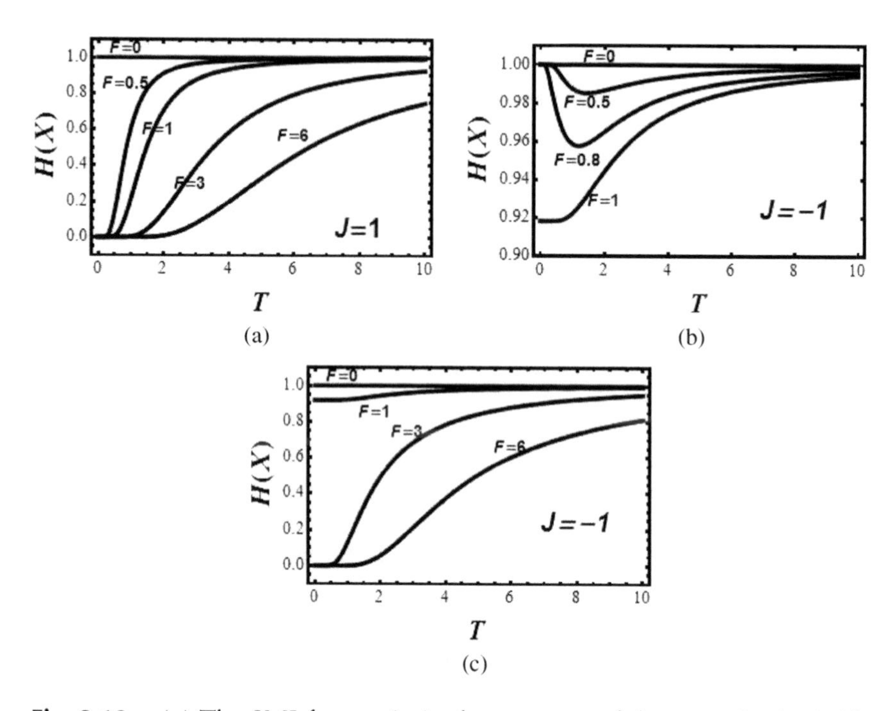

Fig. 3.18. (a) The SMI for a spin in the presence of the second spin (with $J = 1$), as a function of T for different values of F. (b) The SMI for a single spin in the presence of the second spin (with $J = -1$), as a function of T for different values of F. (c) For larger fields.

is a result of having only two states with equal probabilities. For each T, increasing the strength of the field F causes a *decrease* in $H(X)$. The reason is that the field favors the "up" state, and therefore the probability distribution becomes more asymmetric as we increase F. On the other hand, for a fixed value of F, increasing the temperature causes an increase in $H(X)$. This is a result of the "randomization" of the two states at high temperature. Clearly, for any F, at very high temperature the two states will become equally probable and the SMI will tend to 1 bit. A similar behavior is seen for $J = -1$ in Fig. 3.19(b).

Figure 3.20 shows the joint SMI for the two cases $J = 1$ and $J = -1$ as a function of temperature. For $F = 0$, the curve is the same

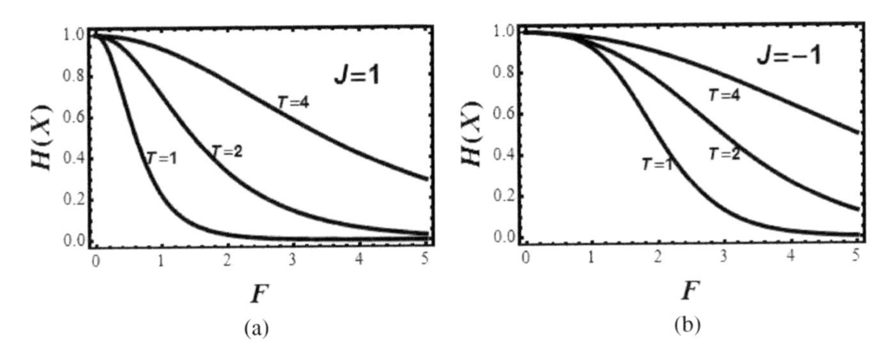

Fig. 3.19. The SMI of a single spin in the presence of field and the presence of the second spin. (a) For $J = 1$ and (b) for $J = -1$.

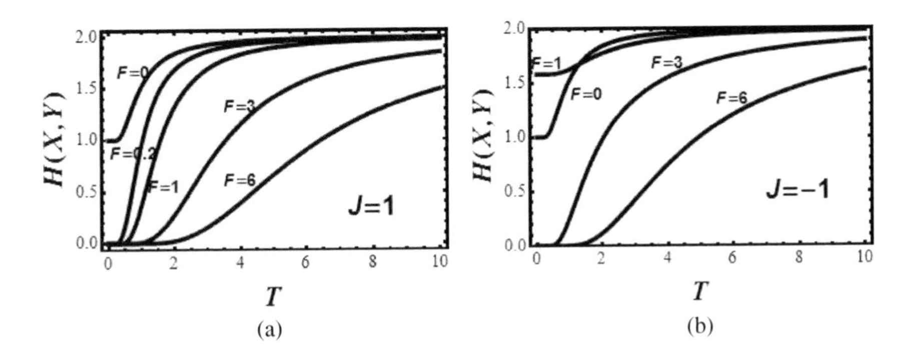

Fig. 3.20. The SMI for the pair of spins: (a) For $J = 1$ and (b) for $J = -1$.

as in Fig. 3.6, i.e. the value of $H(X, Y)$ starts at $H(X, Y) = 1$ at $T = 0$, then reaches $H(X, Y) = 2$ at higher temperatures.

For $J = 1$, we observe that turning on the external field always decreases the value of $H(X, Y)$. This is a result of the fact that for $J = 1$, the interaction between the spins and the interaction of the spins with the field operates in the same direction favoring the $(1, 1)$ state. This means that the uncertainty, with respect to the four states, *decreases* with F, but for each F the uncertainty *increases* with the temperature.

A different behavior occurs for $J = -1$. Here, for small values of $F < 1$, the joint SMI increases with F, but for higher values of F it

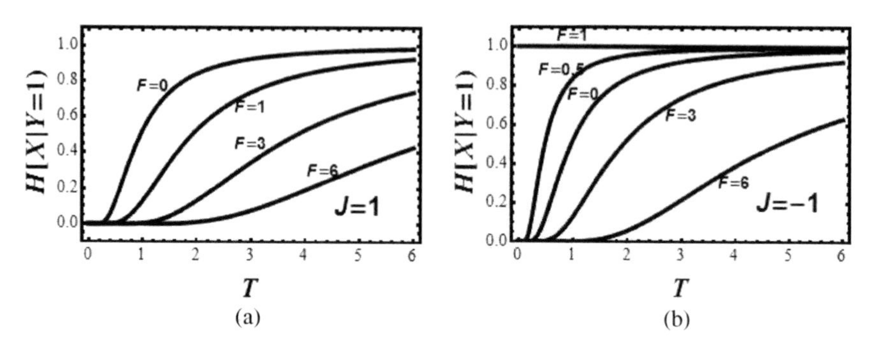

Fig. 3.21. The conditional SMI of one spin, X: (a) For given $y = 1$, and (b) for given $y = 1$.

becomes smaller, and for very large F the value of $H(X, Y)$ reaches zero. At very large fields, there is only one favored state, $(1, 1)$, in both of the cases $J = 1$ and $J = -1$.

Figures 3.21(a) and 3.21(b) show the conditional SMI of X given $Y = 1$ and $Y = -1$, respectively. We see that for $J = 1$, when $Y = 1$, turning on the field enhances the probability of the "up" state, which is also enhanced by the interaction, and hence the conditional information only decreases upon increasing F. A different behavior is observed for $Y = -1$. Here, "turning on" the field initially increases the conditional information. However, when the field is very strong, it becomes the dominant factor in orienting X "up," and the conditional information decreases with the increase of the field.

Figure 3.22 shows the conditional information $H(X|Y)$, which is the average of the two curves in Fig. 2.21. We see that the conditional information always decreases with F. The larger the F, the larger the information on X given Y. A different behavior is observed for the case $J = -1$. This case is left to the reader as an exercise.

Figure 3.23 shows the mutual information $I^{(2)}(X; Y)$ for $J = 1$ and various values of F. Compare the curves in Fig. 3.22 with those in Fig. 3.21. On the other hand, fixing F, the mutual information starts at $I^{(2)}(X; Y) = 0$ at $T = 0$ (except for $F = 0$), then goes through a maximum, and eventually tends to zero at very high temperature.

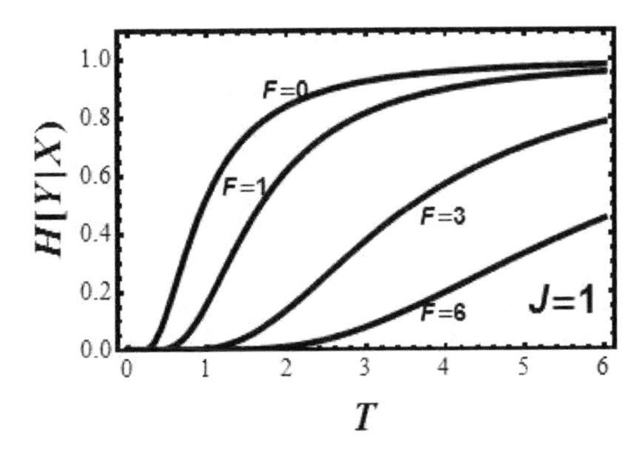

Fig. 3.22. The conditional SMI of the spin X, given that the outcome of Y is known.

For any finite F at $T = 0$, the two spins are oriented in the "up" state. In this case $H(X)$, $H(Y)$, and $H(X, Y)$ are all zeros, and hence also $I^{(2)} = 0$. This is true except for the case $F = 0$, in which case, at $T = 0$, there are *two* equally probable states, "up–up" and "down–down." In this case, knowing X, say "up," determines the orientation of Y.

For $F \neq 0$, the situation is more complicated. Knowing X gives some information on Y, and vice versa. However, when the temperature is very large it tends to randomize the orientations of the two spins, and hence knowing X provides less information on Y. At $T \to \infty$, knowing X does not provide any information on Y, and vice versa.

3.4.3 *Two Interacting Spins in a Nonuniform External Field*

This is a more general case where the potential function is

$$U(x, y) = -J \times (x \times y) - F_1 \times x - F_2 \times y. \tag{3.80}$$

Here, the strength of the external field acting on each spin is different.

The reader is urged to study this system in detail. Calculate the relevant probabilities, the various SMI's, and the mutual

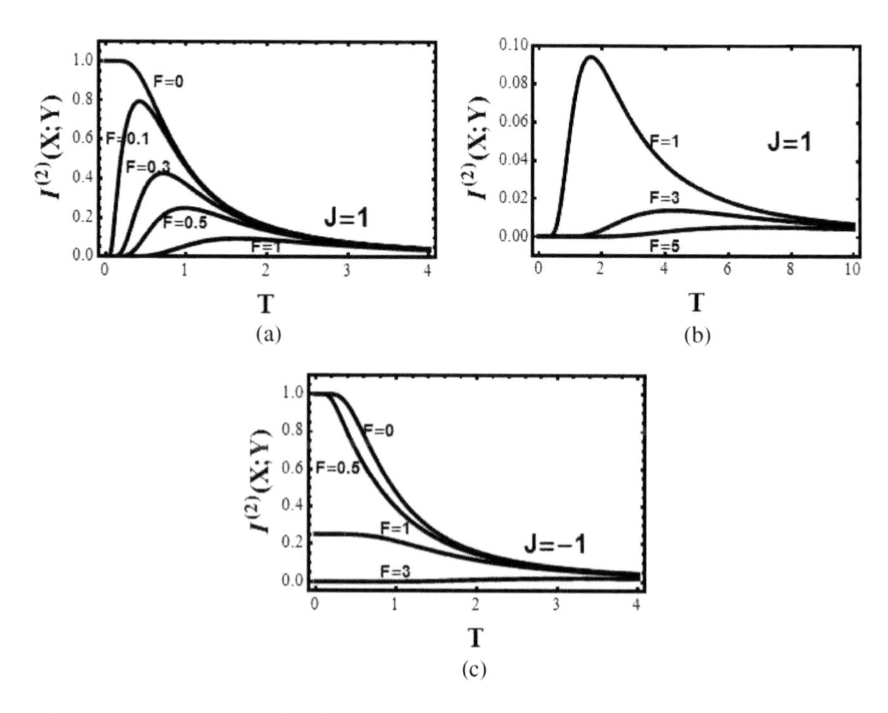

Fig. 3.23. The mutual information for the two spins as a function of T and $J = 1$: (a) For small values of F, (b) for large values of F, and (c) the same as in Fig. 3.22 but for $J = -1$.

information, compare the results with the one in the previous section, and explain why there are such differences between the results.

3.5 An Evolving SMI in a Markov Chain

We recall that a stationary Markov chain is completely characterized by an initial distribution $\boldsymbol{\pi}^{(0)} = (\pi_1^{(0)}, \ldots, \pi_N^{(0)})$, and a transition matrix $\boldsymbol{P}$. The distribution at "time" n is determined by

$$\boldsymbol{\pi}^{(n)} = \boldsymbol{\pi}^{(0)}\boldsymbol{P}^{(n)}. \tag{3.81}$$

Thus, for any sequence of experiments $X_0, X_1, X_2, \ldots, X_n$ forming a stationary Markov chain, the SMI at "time" n defined on the distribution $\boldsymbol{\pi}^{(n)}$ is determined by the initial distribution and the transition matrix. In the following section, we will study a few examples of such an "evolving" SMI.

In Sec. 0.11, we noted that a property equivalent to the Markovian property can be stated as follows: given the "present," the conditional probabilities of the "past" and "future" are independent. We wrote this as

$$p(x_1, x_2, \ldots, x_{k-1}, x_{k+1}, \ldots, x_n | x_k) = p(x_1, x_2, \ldots, x_{k-1} | x_k) \times p(x_{k+2}, \ldots, x_n | x_k). \quad (3.82)$$

On the left hand side, we have the conditional probability of both the "past" and the "future," given the "present." On the right hand side, we have the product of the conditional probabilities, of the "past" given the "present" and the "future" given the "present".

Corresponding to this independence, we can write the conditional SMI

$$H(X_1, X_2, \ldots, X_{k-1}, X_{k+1}, \ldots, X_n | X_k) = H(X_1, X_2, \ldots, X_{k-1} | X_k) + H(X_{k+1}, \ldots, X_n | X_k). \quad (3.83)$$

In words, Eq. (3.83) means that the conditional SMI associated with the joint experiments $(X_1, X_2, \ldots, X_{k-1}, X_{k+1}, \ldots, X_n)$ is the sum of the conditional SMI associated with the joint experiments: $(X_1, X_2, \ldots, X_{k-1})$ and $(X_{k+1}, \ldots, X_n)$, given X_n. Equivalently, we can say that knowing the "presence" makes the information on the two experiments $(X_1, X_2, \ldots, X_{k-1})$ and $(X_{k+1}, \ldots, X_n)$ independent. In terms of mutual information, we can also write the equality

$$I(X_1, X_2, \ldots, X_{k-1}; X_{k+1}, \ldots, X_n | X_k) = 0. \quad (3.84)$$

Note the location of the semicolon separating the two experiments: "past" $(X_1, X_2, \ldots, X_{k-1})$ and "future" $(X_{k+1}, \ldots, X_n)$.

As we have emphasized in Sec. 0.11, the index n does not necessarily signify the "time" at which the nth experiment is carried out. Therefore, we put the words "present," "past," and "future" in inverted commas.

A particular example of the equality (3.84) is the following: for any three consecutive experiments X_k, X_{k+1}, and X_{k+2} from a Markov chain, we have the equality

$$I(X_{k+2}; X_k|X_{k+1}) = 0. \tag{3.85}$$

Since the Markov chain is stationary, the equality (3.85) does not depend on the index k, and hence we can write

$$I(X_3;\ X_1|X_2) = I(X_1;\ X_3|X_2) = 0. \tag{3.86}$$

Exercise 3.5

Prove that the equality (3.86) follows directly from the Markovian property. (See Note 1.)

Another interesting result for a Markov chain is the following:

For any consecutive quadruplet of experiments X_1, X_2, X_3, X_y forming a Markov chain, we have the inequality

$$I(X_1; X_4) \leq I(X_2; X_3). \tag{3.87}$$

This means that the mutual information between two "closer" experiments is always larger than that for the two "farther" experiments.

The proof follows essentially from the definition of the mutual information and the Markovian property.

3.6 The Ehrenfest Model

This model, originally referred to as a "diffusion model," was designed to explain the approach to equilibrium of a thermodynamic system. We bring it up here as an example of a Markov chain. We will see in Chap. 5 that this model is also relevant to the second law of thermodynamics.

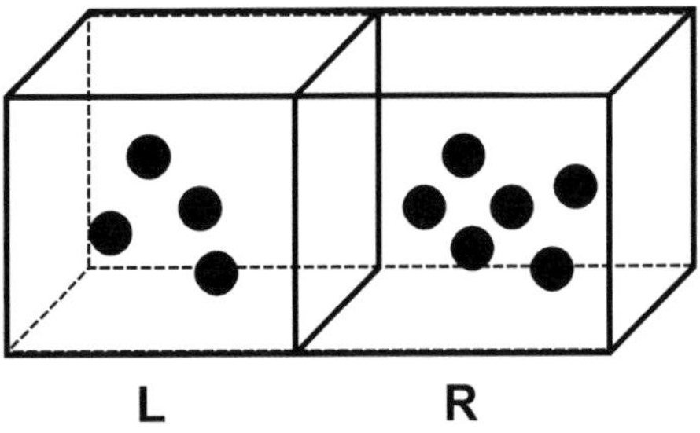

Fig. 3.24. Ten marbles distributed in two compartments, L and R.

Let us assume that we have a total of N marbles, distributed in two compartments L and R (Fig. 3.24) having equal volumes. We define the sequence of experiments, or rv's, $X_1, X_2, \ldots, X_n, \ldots$, such that the event $(X_n = i)$ means that at "time" n the compartment L contains i marbles (and hence the compartment R contains $N - i$ marbles). The number i describes the "state" of the system. We assume that at each "time" n the number of marbles in L can either go up by one unit or down by one unit. We also assume that the transition probabilities are given by

$$P_{ii} = P(X_n = i | X_{n-1} = i) = 0, \tag{3.88}$$

i.e. there is zero probability of staying in the state "i."

The probability of going "up" from the state i to the state $i + 1$ is

$$p_{i,i+1} = P(X_{n+1} = i + 1 | X_n = i) = \frac{N - i}{N} \tag{3.89}$$

and the probability of going "down" from the state i to the state $i - 1$ is

$$P_{i,i-1} = P(X_{n+1} = i - 1 | X_n = i) = \frac{i}{N}. \tag{3.90}$$

Thus, if at "time" n there are i particles in L, there is zero probability that the number of marbles in L will not change in the next step. The probability that the number of marbles in L will *increase* by one is proportional to the fraction of marbles in R. The probability that the number of marbles in L will decrease by one is proportional to the fraction of marbles in L.

The idea is that we choose a specific marble at random. If that marble happens to be in L, it will be transferred to R; if it happens to be in R, it will be transferred to L. Clearly, if we have i marbles in L and $N - i$ marbles in R, and we select a marble at random, the probability of that marble being in L is $\frac{i}{N}$, and the probability that it will be in R is $(N - i)/N$.

Thus, if the selected marble was found in L [with probability $P(\mathrm{L}) = \frac{i}{N}$], it will be transferred to R, and hence the number of particles in L will *decrease*. Similarly, if the selected marble was in R [with probability $P(R) = \frac{N-i}{N}$], it will be transferred to L, and hence the number of marbles in L will *increase*.

The transition matrix, say for $N = 5$, has the general form

$$\boldsymbol{P} = \begin{pmatrix} 0 & 1 & 0 & 0 & 0 \\ 1/5 & 0 & 4/5 & 0 & 0 \\ 0 & 2/5 & 0 & 3/5 & 0 \\ 0 & 0 & 4/5 & 0 & 1/5 \\ 0 & 0 & 0 & 1 & 0 \end{pmatrix} \tag{3.91}$$

Note that all $P_{ii} = 0$, and $P_{0,1} = 1$ (i.e. if there are zero particles in L, then the selected marble must be in R, and the transition of the marble must be from R to L). Similarly, $P_{N,N-1} = 1$ (i.e. if all the marbles are in L, the selected marble must be in L, and hence the number of marbles in L must decrease by one).

We define the distribution vector at "time" n by $\boldsymbol{\pi}^{(n)} = (\pi_0^{(n)}, \pi_1^{(n)}, \ldots, \pi_N^{(n)})$. Here, $\pi_i^{(n)}$ is the probability that at time n there are i marbles in the compartment L. Note that if i is even, then one cannot get back to i in an odd number of steps.

A *stationary distribution* $\boldsymbol{\pi} = (\pi_0, \ldots, \pi_N)$ is defined by

$$\pi = \pi \boldsymbol{P}, \tag{3.92}$$

where on the right hand side we have a product of a row vector and a matrix $\boldsymbol{P}$. Note that $\boldsymbol{\pi}$ is the same vector on both sides of Eq. (3.92).

In terms of the components, π_j is obtained by the scalar product of $\boldsymbol{\pi}$ and the jth column of the matrix $\boldsymbol{P}$:

$$\pi_j = \sum_{j=1}^{N} \pi_i P_{ij} = \pi_{j-1} P_{j-1,j} + \pi_{j+1} P_{j+1,j}. \tag{3.93}$$

This means that we can get to the state j either from the state $j-1$ and adding 1, or from the state $j+1$ and deducting 1. According to the assumptions of the model [see Eqs. (3.88)–(3.90)],

$$\pi_j = \pi_{j-1} \frac{N-j+1}{N} + \pi_{j+1} \frac{j+1}{N}. \tag{3.94}$$

This is valid for $j = 1, 2, \ldots, N-1$ and we also have

$$\pi_0 = \frac{\pi_1}{N}, \tag{3.95}$$

i.e. the state "0" can be obtained only from the state "1," and

$$\pi_N = \frac{\pi_{N-1}}{N}, \tag{3.96}$$

i.e. the state "N" can be obtained only from "N − 1," and the normalization condition is

$$\sum_{j=0}^{N} \pi_j = 1. \tag{3.97}$$

Altogether, we have $N+1$ equations (3.94) and (3.95) for $N+1$ unknowns $(\pi_0, \pi_1, \ldots, \pi_N)$.

We can write Eq. (3.94) as a recursion equation of the form

$$\pi_{j+1} = \left(\pi_j - \pi_{j-1} \frac{N-j+1}{N}\right)\left(\frac{j+1}{N}\right) \tag{3.98}$$

or, equivalently,

$$\pi_j = \left(\pi_{j-1} - \pi_{j-2} \frac{N-j+2}{N}\right)\left(\frac{j}{N}\right). \tag{3.99}$$

From these equations, we can write all the π_j $(j = 1, \ldots, N)$ in terms of π_0:

$$\begin{aligned}
\pi_1 &= N\pi_0, \\
\pi_2 &= (\pi_1 - \pi_0)\frac{N}{2} = \frac{\pi_0 N(N-1)}{2}, \\
\pi_3 &= \left(\pi_2 - \pi_1 \frac{N-1}{N}\right)\frac{N}{3} = \frac{\pi_0 N(N-1)(N-2)}{2 \times 3}, \\
&\vdots \\
\pi_k &= \frac{\pi_0 N(N-1)(N-2)\cdots[N-(k-1)]}{k!} = \binom{N}{k}\pi_0, \\
\pi_N &= \frac{\pi_0 N(N-1)\cdots 2 \times 1}{N!} = \pi_0.
\end{aligned} \tag{3.100}$$

Since π is a distribution, we must have

$$1 = \sum_{k=0}^{N} \pi_k = \sum_{k=0}^{N} \binom{N}{k} \pi_0 = 2^N \pi_0. \tag{3.101}$$

From this equation, we get

$$\pi_0 = \left(\frac{1}{2}\right)^N \tag{3.102}$$

and

$$\pi_k = \binom{N}{k}\left(\frac{1}{2}\right)^N, \tag{3.103}$$

which is the binomial distribution.

Figure 3.25 shows the stationary probability distribution $P(j)$ for $N = 20, 40, 60$ after 100 steps. Note that we have changed the notation from π_j to $P()$. It is clear that the general trend is toward a binomial distribution. In Chap. 5, we will further discuss this experiment in connection with the second law of thermodynamics, where the stationary distribution is identified as the equilibrium distribution.

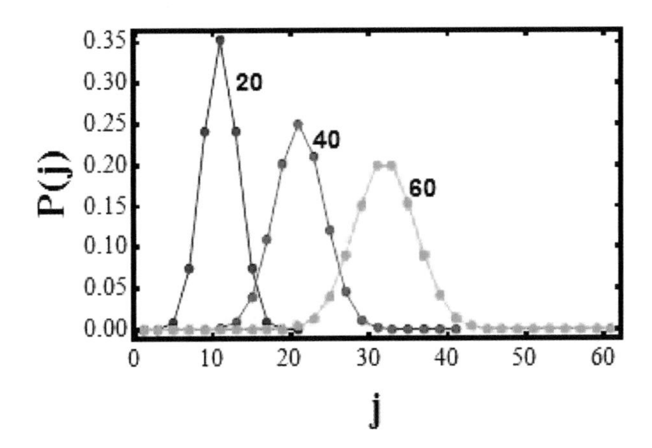

Fig. 3.25. The distribution $p(j)$ for different $N = 20, 40, 60$.

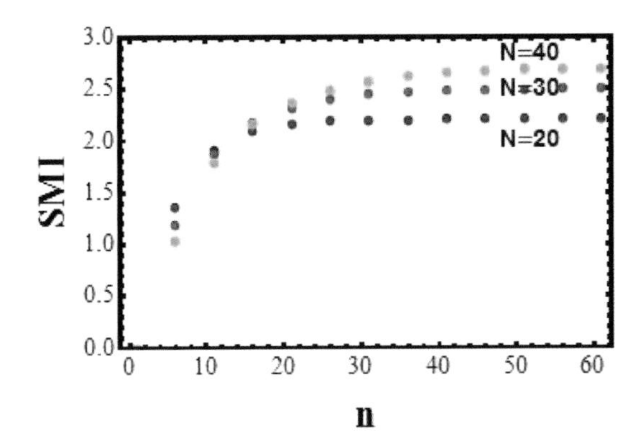

Fig. 3.26. Change in the SMI as a function of the number of steps for different N.

It is clearly seen from Fig. 3.25 that for any N there is a maximum probability at $j \approx N/2$, and the value of the maximum probability *decreases* with N.

Figure 3.26 shows the "evolution" of the SMI with "time" n, defined by

$$H = -\sum_{j=0}^{N} P(j) \log P(j), \tag{3.104}$$

for the cases $N = 20, 30, 40$. The general trend is clear: the SMI initially increases toward a maximum value, and thereafter does not change any further.

A caveat

Do not rush to conclude that the SMI always increases for a series of experiments for which the distribution changes with time, or to conclude that in any Markov process the SMI always increases with "time."

Consider the following Markov process [cf. Eqs. (3.89) and (3.90)]:

$$P_{i,i-1} = \frac{N-i}{N}, \quad P_{i,i+1} = \frac{i}{N}. \tag{3.105}$$

In such a model, starting with any initial distribution, the SMI will *decrease* to zero. One can devise experiments for which an SMI will oscillate with time.

Exercise 3.6

Suppose that we have 20 numbered balls in box A. At each unit of time, the balls numbered 1–10 are transferred from the box they are into another box. Denote by P_5, the probability of finding the ball numbered 5 in box A and $1 - P_5$ in box B. How will the SMI defined on the distribution $(P_5, 1 - P_5)$ change with time?

In Fig. 3.27, we show the result of the simulation of the process according to the following protocol. We start with all the $N = 1000$ marbles in the compartment R (no marbles in the compartment L). Choose a specific marble at random; if it is in L it will be transferred to R, and if it is in R it will be transferred to L. In Fig. 3.27(a), we show the change in the number of marbles in L as a function of the number of steps, n. We see that at about 2000 steps the number of marbles in L tends to about 500. Figure 3.27(b) shows the evolution of the SMI for the same experiment. Here, the SMI tends to 1 as n increases (i.e. one bit per marble).

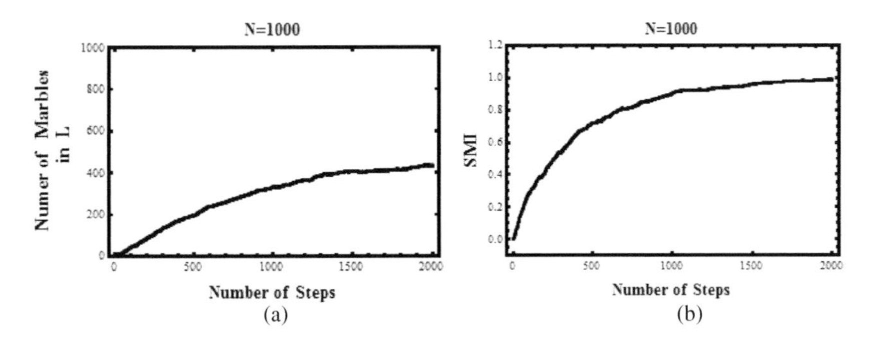

Fig. 3.27. Simulation of expansion of 1000 marbles: (a) The change in the number of marbles in compartments L and (b) the change of the SMI per particle.

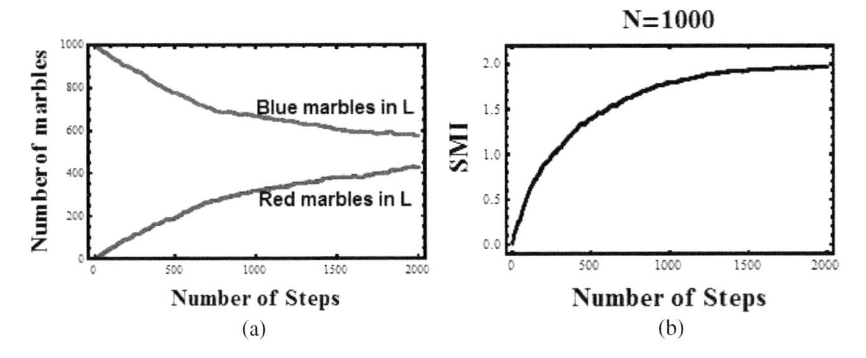

Fig. 3.28. Simulation of mixing of 1000 red and 1000 blue marbles: (a) The change in the number of marbles in each compartments, and (b) the change of the SMI.

Figure 3.28 shows similar data for the experiment of *mixing* red and blue marbles. Initially, all the red marbles are in the compartment R (i.e. no red marbles in L) and all the blue marbles are in the compartment L. As can be seen, after a large number of steps both the red and the blue marbles tend to "expand," to occupy the two compartments, exactly as if we had carried out the experiment with the red and the blue marbles separately. Figure 3.28(b) shows the evolution of the SMI as a function of the number of steps. The curve tends to 2 (again one bit per marble, and altogether there are 1000 marbles of each color).

3.7 An Evolving SMI in a Two-State Markov Chain

In this section, we study a very simple case of a Markov process for which the SMI can increase, decrease, or even increase and decrease with "time." The experiment has two possible outcomes; these could be either *head* or *tail* in tossing a coin, or a sequence of zeros or ones. We denote the initial distribution by $\boldsymbol{\pi}^{(0)} = (\pi_1^{(0)}, \pi_2^{(0)})$ and the transition matrix

$$P = \begin{pmatrix} 1-\alpha & \alpha \\ \beta & 1-\beta \end{pmatrix}. \tag{3.106}$$

Here, $0 \leq \alpha \leq 1$ is the transition probability for the change of states $1 \to 2$, and $0 \leq \beta \leq 1$ is the transition probability for the change $2 \to 1$, and $1-\alpha$ and $1-\beta$ are the conditional probabilities of staying in the same state, i.e. P_{11} and P_{22}, respectively.

Clearly, this is a stochastic matrix; the sum of terms in each row is 1.

Before we study the general case, we examine some special cases:

(a) The case $\alpha = \beta = 0$

In this case, the transition matrix is

$$P = \begin{pmatrix} 1 & 0 \\ 0 & 1 \end{pmatrix}. \tag{3.107}$$

This means that in any transition the distribution does not change, i.e.

$$(\pi_1, \pi_2)\begin{pmatrix} 1 & 0 \\ 0 & 1 \end{pmatrix} = (\pi_1, \pi_2). \tag{3.108}$$

Also, the SMI associated with the distribution does not change.

(b) The case $\alpha = \beta = 1$

In this case, the transition matrix is

$$P = \begin{pmatrix} 0 & 1 \\ 1 & 0 \end{pmatrix}. \tag{3.109}$$

Clearly in this case, at each step the states change with probability 1, from $1 \to 2$ and from $2 \to 1$. In general, we have

$$(\pi_1, \pi_2) \begin{pmatrix} 0 & 1 \\ 1 & 0 \end{pmatrix} = (\pi_2, \pi_1). \tag{3.110}$$

This is also a deterministic process, as in case (a). Both case (a) and case (b) may be referred to as deterministic Markov chains; in the first case, the distribution (π_1, π_2) is preserved, whereas in the second π_1 and π_2 exchange at each step. In both cases, the SMI does not change.

(c) The case $\alpha + \beta = 1$

In this case, the transition matrix is

$$P = \begin{pmatrix} \beta & \alpha \\ \beta & \alpha \end{pmatrix}. \tag{3.111}$$

It is easy to see that all powers of P are equal to P, for instance,

$$P^2 = \begin{pmatrix} \beta & \alpha \\ \beta & \alpha \end{pmatrix} \begin{pmatrix} \beta & \alpha \\ \beta & \alpha \end{pmatrix} = \begin{pmatrix} \beta & \alpha \\ \beta & \alpha \end{pmatrix}, \tag{3.112}$$

and by generalization

$$P^n = P. \tag{3.113}$$

If the initial vector is $\boldsymbol{\pi}^{(0)} = (\pi_1^{(0)}, \pi_2^{(0)})$, we get after the first step

$$\boldsymbol{\pi}^{(0)} P = \left(\pi_1^{(0)}, \pi_2^{(0)}\right) \begin{pmatrix} \beta & \alpha \\ \beta & \alpha \end{pmatrix} = (\beta, \alpha). \tag{3.114}$$

Thus, from *any* initial distribution $\boldsymbol{\pi}^{(0)}$, the next distribution will be (β, α), and it will remain unchanged in all other steps. We can say that after the first step the system does not "remember" the initial distribution; the distribution at any step $n (n \neq 0)$ is independent of the initial distribution.

(d) The general case $0 < \alpha + \beta < 2$

We now prove an important theorem: the transition matrix P^n tends to a constant matrix when $n \to \infty$. The mathematical details are shown in Note 3. The result is

$$\boldsymbol{P}^n \to \begin{pmatrix} \frac{\beta}{\alpha+\beta} & \frac{\alpha}{\alpha+\beta} \\ \frac{\beta}{\alpha+\beta} & \frac{\alpha}{\alpha+\beta} \end{pmatrix}, \quad \text{for } n \to \infty. \tag{3.115}$$

This means that starting with any initial vector $\boldsymbol{\pi}^{(0)}$, after a large number of steps, n, we get

$$\begin{aligned} \boldsymbol{\pi}^{(0)} \boldsymbol{P}^n &\to \frac{\left(\pi_1^{(0)}, \pi_2^{(0)}\right)}{(\alpha+\beta)} \begin{pmatrix} \beta & \alpha \\ \beta & \alpha \end{pmatrix} = \frac{(\beta, \alpha)}{(\alpha+\beta)} \\ &= \left(\frac{\beta}{(\alpha+\beta)}, \frac{\alpha}{(\alpha+\beta)} \right) = \boldsymbol{\pi}^{(\infty)}. \end{aligned} \tag{3.116}$$

Note that $\pi_1^{(0)} + \pi_2^{(0)} = \pi_1^{(\infty)} + \pi_2^{(\infty)} = 1$.

Regarding the evolution of the SMI associated with this experiment, we can conclude that starting with any stochastic matrix of the form (3.106), and any initial distribution $\boldsymbol{\pi}^{(0)}$, the SMI will evolve into a constant, determined by the limiting distribution

$$\boldsymbol{\pi}^{(\infty)} = \lim_{n \to \infty} \boldsymbol{\pi}^{(0)} \boldsymbol{P}^n = \left(\frac{\beta}{\alpha+\beta}, \frac{\alpha}{\alpha+\beta} \right). \tag{3.117}$$

Since this is a two-state experiment, the SMI, at any "time" n, has the form

$$H(p) = -p \log p - (1-p) \log (1-p), \tag{3.118}$$

where $p = \pi_1^{(n)}$ is a function of "time." Since $\pi_1^{(n)}$ changes with "time," the SMI will also change with "time," and it will reach a limiting value after many steps.

Thus, choosing a matrix of the form (3.106) with any α and β such that $0 < \alpha < 1, 0 < \beta < 1$, and $0 < \alpha < \beta < 2$, the limiting SMI would be $H\left(\frac{\beta}{\alpha+\beta}\right)$.

For instance, if we choose $\alpha = \beta = 1/2$, then from any initial distribution $\pi^{(0)}$ the SMI of the system will *increase* toward 1, in one step.

For example,

$$\pi^{(0)} = (0.2, 0.8), \quad H(0.2) = 072;$$

we get in one step

$$\pi^{(1)} = \pi^{(0)} P = (0.5, 0.5). \tag{3.119}$$

Thus, starting at point C in Fig. 3.29, the value of the SMI will increase toward 1 along the green line in the figure.

On the other hand, if we choose any other pair of values of α and β, then the SMI could either *increase* or *decrease* toward the limiting value of the SMI; see Fig. 3.29.

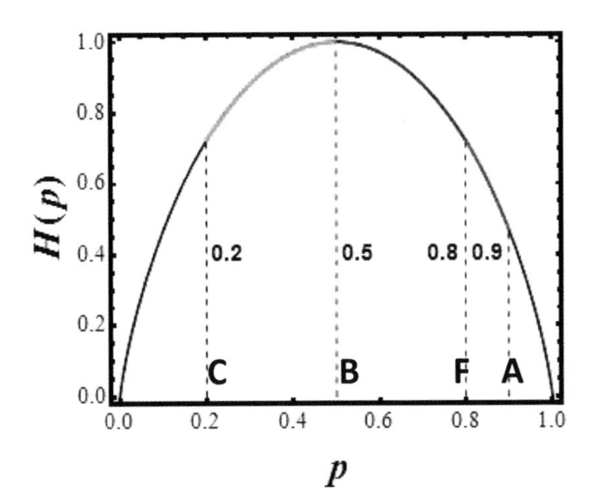

Fig. 3.29. The SMI as a function of the p, with a final $p = 0.8$ (point F), for different initial distributions, A, B, and C.

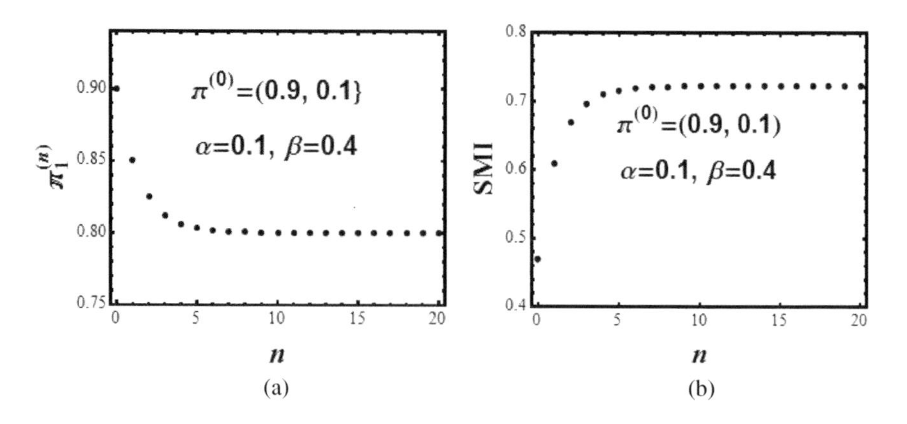

Fig. 3.30. The case of the SMI increasing with "time."

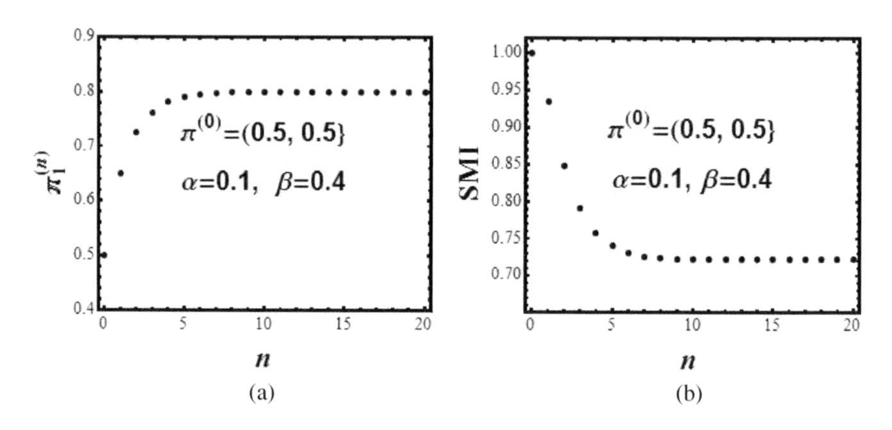

Fig. 3.31. The case of the SMI decreasing with "time."

Figure 3.30 shows an example of an *increasing* SMI from the initial value of SMI $= 0.47$, corresponding to $\pi^{(0)} = (0.1, 0.9)$ to the final value of SMI $= 0.97$, corresponding to the stationary distribution $\pi^{(\infty)} = (0.4, 0.6)$.

Figure 3.31 shows an example of a *decreasing* SMI from the initial value of SMI $= 1, 0$ corresponding to $\pi^{(0)} = (0.5, 0.5)$ to the final value of SMI $= 0.72$, corresponding to the stationary distribution $\pi^{(\infty)}$. Here, the SMI decreases along the blue line in Fig. 3.29.

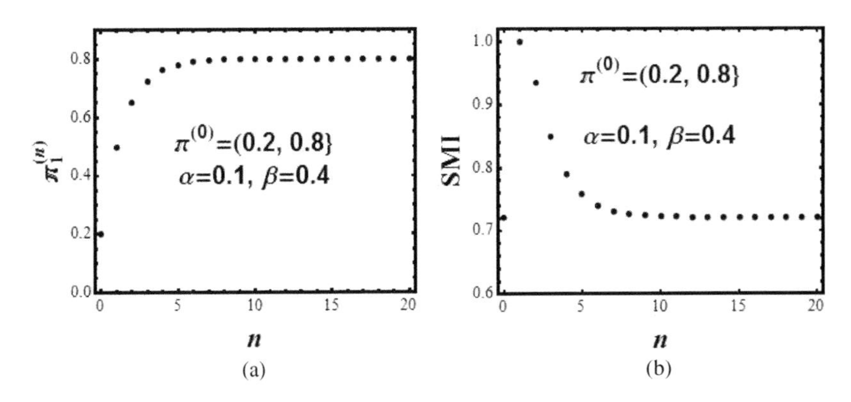

Fig. 3.32. The case of the SMI first increasing then decreasing with "time."

Clearly, whenever we start with a distribution for which $1/2 < p < 1$, the SMI will either increase or decrease toward the final limiting value of 0.72. An interesting behavior is shown when we choose $0 < p < 1/2$. Figure 3.32 shows a case where the SMI initially increases as a function of n, but then decreases toward the final value of SMI = 0.72. Here, we start with the initial distribution of $\pi^{(0)} = (0.2, 0.8)$. Note that at this initial distribution the SMI is the same as for the limiting distribution, i.e. SMI = 0.72. However, as we can see, in this case the SMI will first "climb" on the SMI curve in Fig. 3.29, along the green line, from the initial value of about 0.72 toward the maximum value of 1, and then *decrease* to the limiting value of 0.72, along the blue line.

Exercise 3.7

In all the illustrations above, we assumed that the Markov chain is stationary. However, one can also study nonstationary Markov chains where the transition matrix is a function of "time." In such a case, we can also get oscillatory behavior of the SMI. Consider, for example, the case where α and β, in the matrix $\boldsymbol{P}$ defined in Eq. (3.106), are functions of n, such as

$$\alpha(n) = 0.1 \cos^2(n),$$
$$\beta(n) = 0.4 \sin^2(n).$$

In such a case, the distribution at "time" n will be given by

$$\pi^{(n)} = \pi^{(0)} \prod_{i=1}^{n} P(i),$$

where $P(i)$ is the transition matrix at "time" i.

Show that by choosing the initial vector $\pi^{(0)} = (0.5, 0.5)$, one gets oscillation in the values of SMI. (See Note 4.)

In Chap. 5, we shall see other examples of macroscopic systems for which the SMI can either increase or decrease with "time."

3.8 A One-Dimensional Lattice of Spins

The final example we study in this chapter is a 1D system of spins. This model is an example of a Markov chain where the index n does not signify the "time" of the experiment, but the location of the spin on lattice points. It is also a generalization of the spin case discussed in Chap. 2.

The system is a linear lattice of equidistant points, at each of which a spin is located. For simplicity, we assume that each spin can be in either of the two orientations, "up" and "down."

The "experiment," or the rv X_i, signifies the state of the spin at the location i. We will be interested in the probabilities of events such as $P(X_i = x_i, X_{i+1} = x_{i+1}, \ldots, X_{i+n} = x_{i+n})$, where x_i can have the two values $(+1)$ for the "up" state and (-1) for the "down" state; see Fig. 3.33. All the probabilities relevant to this system can be obtained from a well-known procedure in statistical mechanics. A brief outline of the theoretical background is provided in App. F.

Fig. 3.33. The 1D spin system.

All the probabilities as well as the SMI for this system are determined by the interaction potential function as defined in Chap. 2:

$$U(x_1, x_2) = -J(x_1 \times x_2) - Fx_1 - Fx_2, \qquad (3.120)$$

where J is a measure of the strength of the interaction between two consecutive spins, and F is the strength of the field which favors the "up," or the (+ 1) state relative to the "down" or the (−1) state.

3.8.1 *Singlet Probabilities*

Figure 3.34 shows the probability of the "up" state, i.e. (+1) of a single spin. Because of the stationary Markov chain property of all the probability distributions, the singlet (as well as the pair) probabilities are independent of the location of the spin in the system.

Figure 3.34(a) shows the probability $P(1)$ for the case $J = 0$ (no pair interactions), as a function of the temperature T, at different values of F.

As expected, for $F > 0$, the "up" state is favored, and therefore at low temperature $P(1) \approx 1$, i.e. the "up" state has probability 1. Similarly, for $F < 0$, the "down" state is favored, and therefore the probability of the "up" state in this case starts at $P(1) \approx 1$ at $T \to 0$. In all cases, increasing the temperature causes gradual randomization, and eventually at very high temperatures all the curves

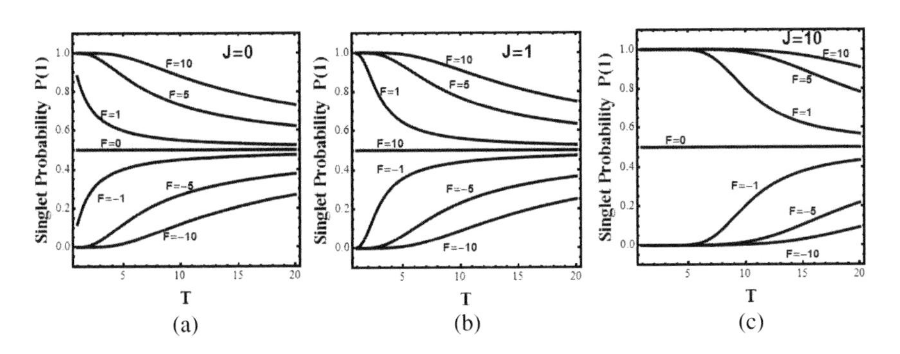

Fig. 3.34. The singlet probabilities for the 1D spin system.

converge to $P(1) \to 1/2$, i.e. equal probability for the two states of the single spin.

Figures 3.34(b) and 3.34(c) show similar data in the presence of interactions with $J = 1$ and $J = 10$. Since $J > 0$ means favoring the parallel states of a pair of spins (either "up–up" or "down–down"), the interaction and the field effects operate in the "same direction," i.e. favoring the "up" state. Thus, the general form of the curves is similar to the case of $J = 0$, except that the tendency for complete randomization is slower as we increase the value of J. In all cases, the eventual value of $P(1)$ at $T \to \infty$ will be 0.5.

Figure 3.35 compares the effect of positive and negative values of J on the singlet probabilities for a fixed value of F. The case of $J > 0$ is clear; the larger the value of J, the slower the decrease of $P(1)$ as a function of T. For negative values of J [Fig. 3.35(b)], the behavior is quite different. In this case, there is a conflict between the effect of the field and the effect of the interaction. This conflicting effect has already been discussed in Secs. 2.12 and 3.4. Briefly, $F > 0$ favors the "up" state. This is true for each spin. When we focus on a specific spin,

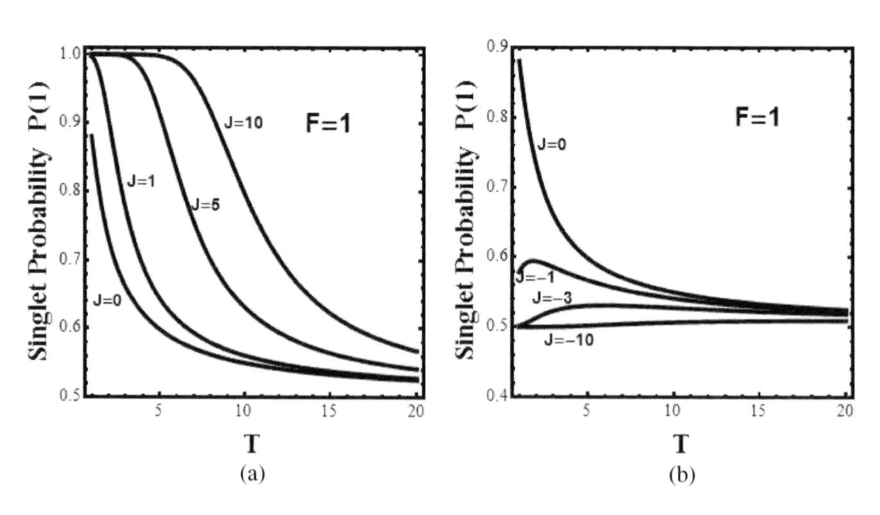

Fig. 3.35. The singlet probabilities for the 1D spin system: (a) For positive values of J, and (b) for negative values of J.

we have two effects. The field favors "up," but the negative J favors the "down" state on the observed spin caused by the neighboring spins. These two effects produce the maximum of $P(1)$ as a function of T. When the interaction parameter is very large and negative, it almost "neutralizes" the effect of the field and the probability of the two states becomes equal. This conflict has been discussed in Sec. 3.4.

3.8.2 *Pair Probabilities*

There are many probabilities for the pair probabilities of the four states "up–up," "up–down," "down–up," and "down–down" at various values of F (positive and negative), and J (positive and negative). We present only a small sample of results. The reader is urged to examine these results carefully as well as calculate other results. We will continue using J for the interaction parameter, and j for the distance between two spins; $j = 1$ for nearest neighbors, $j = 2$ for next nearest neighbors, etc.

Figure 3.36(a) shows $P(1, 1)$ for a *nearest neighbor* pair of spins (i.e. $J = 1$) for $J = 1$ and positive values of the field F. Since J and F operate in the same direction, favoring the "up–up" states, the results in Fig. 3.36(a) are obvious. Figure 3.36(b) shows similar data for $P(1, -1)$. Here, all the curves start at $P(1, -1) = 0$ at low temperatures, and as the temperature increases all the curves converge

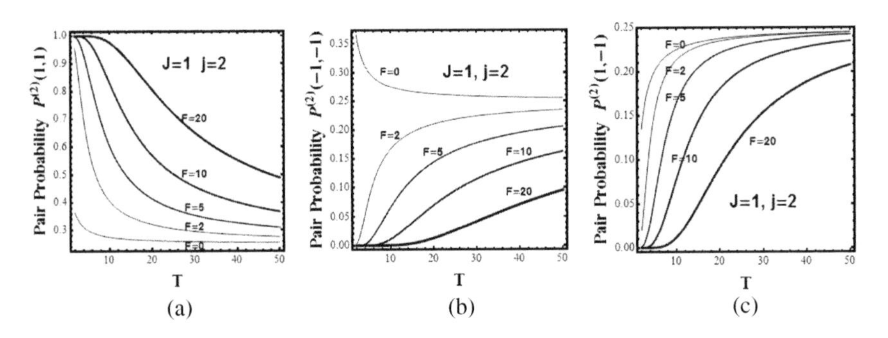

Fig. 3.36. Some pair probabilities for the 1D spin system.

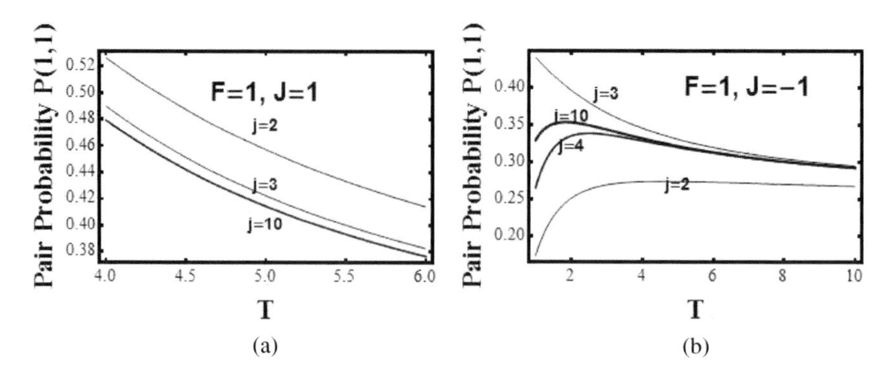

Fig. 3.37. Some pair probabilities for the 1D spin system for several values of the distance between the spins.

to 0.25 (i.e. equal probabilities for the four states of the pair of spins).

A slightly different behavior is shown by $P(-1,-1)$. Here, the case of $F = 0$ is the same as in Fig. 3.36(a), but for any other $F > 0$ we find that $P(-1,-1)$ starts with the value of zero at $T = 0$, because the field favors the "up–up" states and disfavors the "down–down" states. As the temperature increases, all the curves converge to 0.25, i.e. total randomization of the four states.

Figure 3.37 shows the probability $P(1,1)$ as a function of T for the case $F = 1, J = 1$ for different values of j. $j = 2$ means second nearest neighbors, $j = 3$ third nearest neighbors, and $j = 10$ tenth nearest neighbors, or two spins separated by nine intermediate spins.

Here, we see that the pair probability decreases with the "distance" between the two spins. For $j = 5$, and larger, there is almost no further change in the curve. At this distance the two spins become independent. See below for the curves for the correlation between pairs of spins. A different behavior is shown in Fig. 3.37(b), where $F = 1$ and $J = -1$.

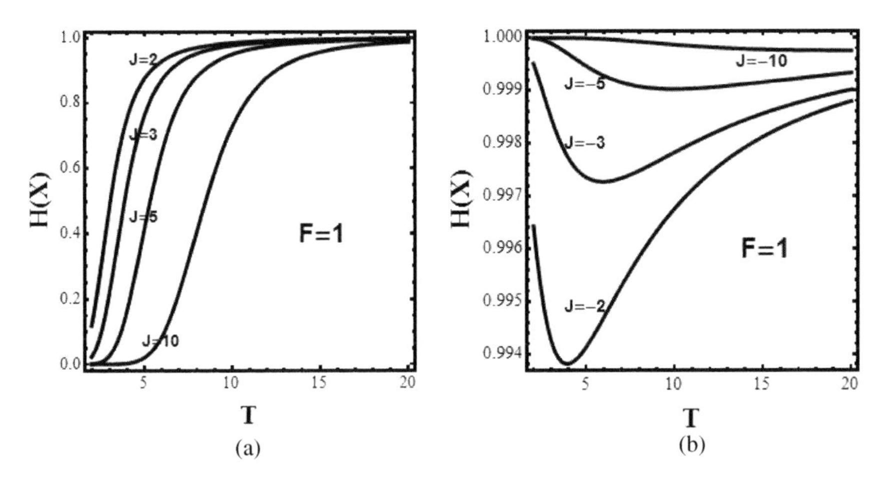

Fig. 3.38. The singlet SMI for the 1D spin system: (a) For positive values of J, and (b) for negative values of J.

3.8.3 *The SMI for Single and Pairs of Spins*

We start with examination of the SMI for a single spin, $H(X)$. Figure 3.38(a) shows $H(X)$ as a function of T for a fixed value of the field $F = 1$ and different positive values of the interaction parameter J. Note that the general trend is one of starting at $H(X) = 0$ at $T \to 0$. This is a result of the dominating effect of the field favoring the "up" state; the probability of the spin to be in this state is nearly 1; hence, the SMI is nearly zero. At very high temperatures, the randomization effect of the temperature overcomes the ordering effect of the field, and therefore the SMI approaches 1. As expected, since the field and the interaction operate in the same direction toward favoring "up", the larger J is, the slower the tendency toward total randomization will be this effect more pronounced the larger the value of $F > 0$ (not shown).

Figure 3.38(b) shows the results for a fixed field $F = 1$, but negative $J < 0$. As we have seen several times earlier, the conflicting effects of the field and the interaction produce maxima or minima in the probability curves. Here, we see that for small negative $J = -2$, the SMI initially decreases with temperature, meaning that the dominating effect is the effect of the field which favors the "up" state. As we increase the absolute value of J, the interaction effect almost cancels out the effect of the field.

At $J = -10$, we see that the curve is almost equal to 1 at all temperatures. This means that the ordering effect of the field is negligible.

Note also that exactly the same curves are obtained for $F = -1$. The reason is that the SMI is defined symmetrically with respect to $P(1)$ and $P(-1)$. Therefore, what we observe for a positive field favoring "up" will be true for negative field favoring "down," and hence the SMI would not change upon changing the sign of F.

Figure 3.39 shows the SMI for a pair of spins for the case $J = 1$ and $F = 1$. We see that the effect of the "distance" (j) is very small. The general trend for large T is clear: as the temperature increases,

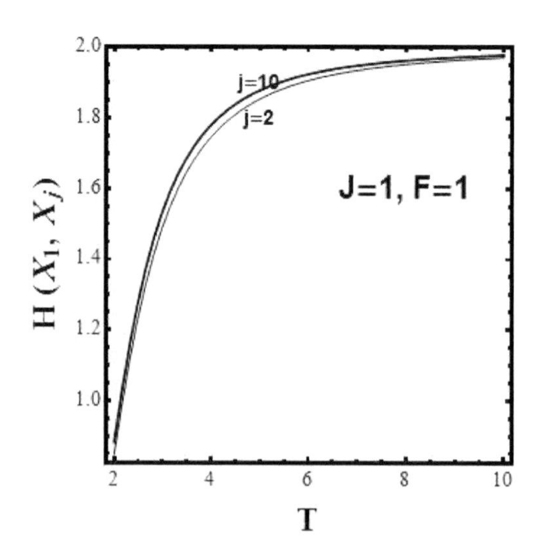

Fig. 3.39. The SMI for pairs of spins at different distance: $j = 2$ and $j = 10$.

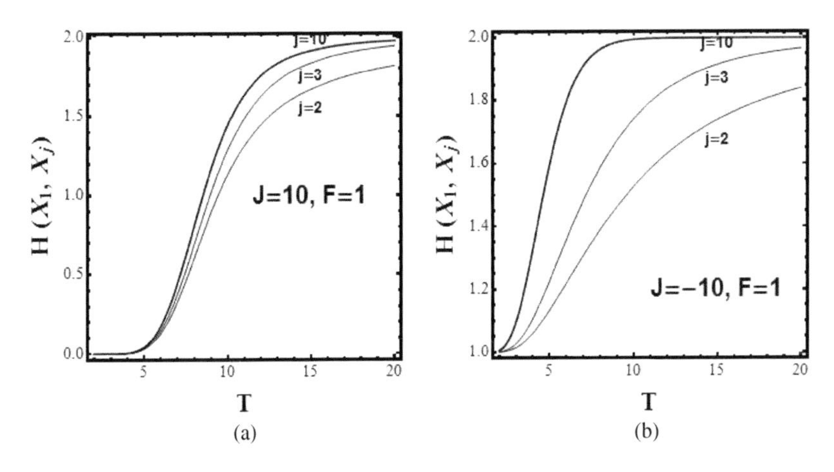

Fig. 3.40. The SMI for pairs of spins at different distance, but with strong spin–spin interactions.

$H(X_1, X_j)$ tends to the value of 2, which is due to the complete randomization of the four states.

Figure 3.40 shows the same SMI for a larger value of the interaction parameter J. When $J = 10$ [Fig. 3.40(a)], we see that all curves start with $H = 0$ at $T \to 0$, and end up with $H = 2$ at $T \to \infty$. This is clear because the field and the interaction affect the spins in the same direction favoring "up." Therefore, at very low temperatures, only one state, "up–up" is favored, and hence $H = 0$. Note also that the larger the distance between the spins, the quicker the curve tends to $H = 2$ at high temperatures.

For the case $J = -10$ [Fig. 3.40(b)], we see similar behavior, but here the curve starts with $H = 1$ at very low temperatures [cf. Fig. 3.40(a)].

This is the result of the conflicting effects of J and F. Thus, when J is very large and negative, it nullifies the effect of the field (which favors "up"). Therefore, at low temperatures, there are only two equally probable states: "up–down" and "down–up." Note also that the results are the same when we change from $F = 1$ to $F = -1$.

Explain why.

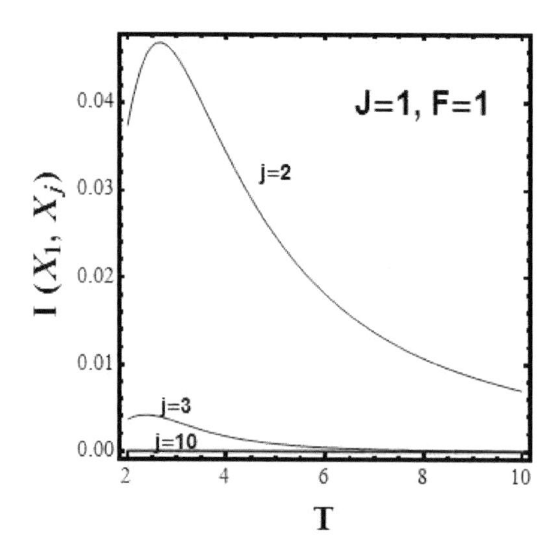

Fig. 3.41. The mutual information for pairs of spins at different distance j.

3.8.4 *Mutual Information*

Finally, we present some values of the mutual information (MI) for pairs of spins at different distances.

Recall that the MI is defined by

$$\begin{aligned} I(X_1; X_j) &= H(X_1) + H(X_j) - H(X_1, X_j) \\ &= 2H(X) - H(X_1, X_j). \end{aligned} \tag{3.121}$$

Note that $H(X)$ is independent of the index i.

Figure 3.41 shows $I(X_1; X_j)$ as a function of T for the case $J = 1$ and $F = 1$, and three values of j. We see that the MI always *decreases* with j, and at very high temperatures all the curves tend to zero, meaning independence of X_1 and X_j.

Figure 3.42(a) shows the MI for the case $J = 10$ and $F = 1$. We see that the MI is much larger than for the case shown in Fig. 3.41. The reason is simple: larger MI means larger correlations between X_1 and X_j. Note also that for any fixed T the MI *decreases* with j. This

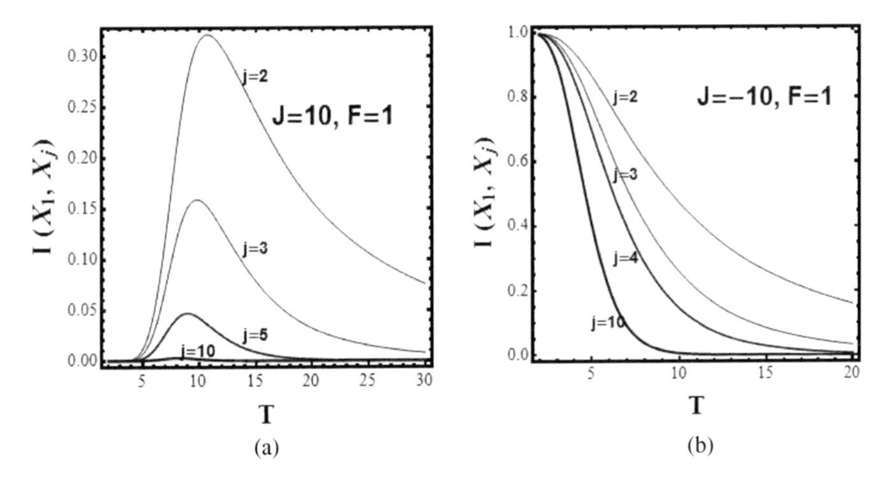

Fig. 3.42. The mutual information for pairs of spins at different distance with strong spin–spin interaction.

means that the larger the distance between "1" and "j," the smaller the correlation. For $J = 10$, there is almost no correlation and the MI is zero for distances $j \geq 10$. Also, note that at higher temperatures all the curves tend to zero — indicating no correlation between X_1 and X_j.

An interesting feature of this case is that also at $T \to 0$ we have zero MI. The reason is that, in this case, the interaction and the field operate to favor the "up" state. As we have seen in Fig. 3.36, the pair probability $P(1, 1)$ and the singlet probability are 1 at $T \to 0$. Therefore, the correlation $g(1, 1)$ is also 1, and the average $\log g$ is 0, i.e. no MI.

A different behavior is shown for the case of $J = -10$ and $F = 1$; see Fig. 3.42(b). Here, there is a conflict between the interaction and the field. As in the previous case, all curves converge to zero at high temperatures. Also, for any fixed T, the MI decreases with increase in the distance j. However, at very low temperatures, the MI starts with the value of 1. The explanation can be drawn from Figs. 3.38

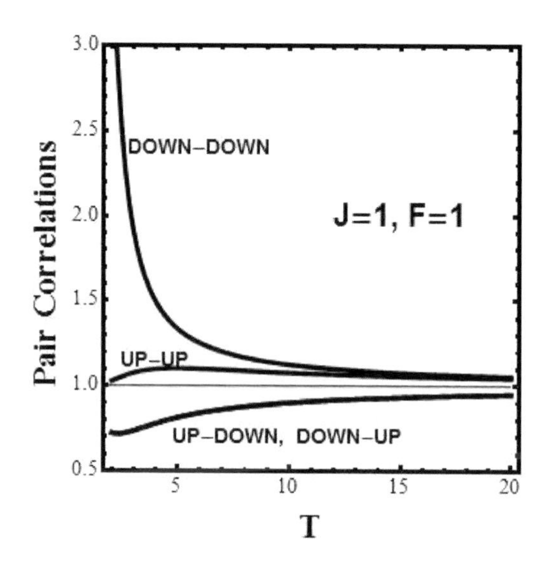

Fig. 3.43. The pair correlations for the different configurations of the pairs of spins.

and 3.40, for $H(x)$ and $H(X_1; X_j)$, which in this case gives

$$
\begin{aligned}
I(X_1; X_j) &= 2H(X) - H(X_1, X_j) \\
&= 2 \times 1 - 1 = 1.
\end{aligned}
\tag{3.122}
$$

Recall that the MI is always positive, although the logarithm of the correlation function can be either positive or negative. We demonstrate this effect by showing the three correlations for the cases "up–up," "down–down," and "up–down." In Fig. 3.43, we see that $g(1, 1)$ is larger than 1. $g(-1, -1)$ is also larger than 1, but $g(1, -1) = g(-1, 1)$ are smaller than 1. This means that two correlations are positive and two negative. The average of the logarithm of the four correlations must be positive. The reason is that, in this case, the probability of the $(1, 1)$ and $(-1, -1)$ states is larger than that of the $(-1, 1)$ and $(1, -1)$ states. Therefore, the average of the logarithm of g will be positive.

3.9 Summary of Chapter 3

In this chapter, we have presented the two important quantities: the conditional and the mutual information. We also studied a few systems for which we can calculate all the relevant probabilities, and hence the conditional and the mutual information too. The reader is urged to examine the data in the various figures, and also other cases not considered in these figures.

4

Multivariate Mutual Information

In Chapter 3, we defined the mutual information (MI) and we saw several ways of interpreting this quantity. In this chapter, we generalize the concept of the MI to any number of experiments or rv's. Some authors decline to define multivariate MI; some even claim that such definition is "illegal." The reason for this is that the MI may be generalized in different ways, and it is not clear what would be the most "natural" or most "useful" generalization of the MI.

In this chapter, we present two possible definitions of multivariate MI, and the relationship between them. We then discuss a few examples of these quantities.

4.1 Multivariate MI Based on Total Correlations

The simplest and perhaps the most "natural" definition of multivariate MI is based on the interpretation of the MI as a measure of the *average correlation* between two rv's (see Sec. 3.3). The MI for two experiments, or rv's, is defined by

$$I(X_1;X_2) = \sum_{x_1,x_2} p(x_1,x_2) \log \frac{p(x_1,x_2)}{p(x_1)\,p(x_2)}$$
$$= \sum_{x_1,x_2} p(x_1,x_2) \log\left[g(x_1,x_2)\right]. \quad (4.1)$$

Here, $g(x_1,x_2)$ is the correlation between the event $X_1 = x_1$ and the event $X_2 = x_2$. When the two events are independent, $g(x_1,x_2) = 1$, we say that the two events are *positively correlated* when $g(x_1,x_2) > 1$ (i.e. $\log[g(x_1,x_2)] > 0$). We say that the two events are *negatively correlated* when $g(x_1,x_2) < 1$ (i.e. $\log[g(x_1,x_2)] < 0$). We see from Eq. (4.1) that $I(X_1;X_2)$ is a measure of the average of the logarithm of the correlation between the two rv's.

The generalization of the quantity in Eq. (4.1) for any number of rv's is straightforward. We define the correlation function for the events $(X_1 = x_1), \ldots, (X_n = x_n)$ by

$$g(x_1,x_2,\ldots,x_n) = \frac{p(x_1,x_2,\ldots,x_n)}{\prod_{i=1}^{n} p(x_i)}. \quad (4.2)$$

Here $p(x_1,\ldots,x_n)$ is the joint probability of finding the events $X_1 = x_1, \ldots, X_n = x_n$. With this definition of the correlation function, we define the *total mutual information* by

$$TI(X_1;\ldots;X_n) = \sum_{x_1,\ldots,x_n} p(x_1,\ldots,x_n) \log\left[g(x_1,\ldots,x_n)\right]. \quad (4.3)$$

As in Eq. (4.1), the quantity *TI* is an average of the logarithm of the correlation function between the n events $(X_1 = x_1), \ldots, (X_n = x_n)$. The quantity $g(x_1,\ldots,x_n)$ is sometimes referred to as the *total correlation*. If all the n rv's are independent, then the total correlation function is unity, i.e. $p(x_1,\ldots,x_n) = \prod_{i=1}^{n} p(x_i)$, and hence $g(x_1,\ldots,x_n) = 1$.

Note that *total* independence does not imply independence between pairs, triplets, etc., of rv's. For specific examples, see Ben-Naim (2015b).

Clearly, the *TI* is always nonnegative. This follows from the inequality (3.55). It is zero if and only if all the n rv's are independent.

4.2 Multivariate MI Based on Conditional Information

The second generalization of the MI is due to Fano (1961). It starts from the identity

$$I(X_1;X_2) = H(X_1) - H(X_1|X_2). \tag{4.4}$$

As we noted in Sec. 3.3, this quantity is always positive. This means that having information on X_2 can only reduce the uncertainty about X_1. We now use the two identities from Sec. 3.3

$$I(X_1;X_1) = H(X_1), \tag{4.5}$$

$$I(X_1;X_1|X_2) = H(X_1|X_2). \tag{4.6}$$

The first equality, (4.5), means that the self-information, i.e. the MI between X_1 and X_1, is the same as the SMI of X_1. The same identity holds for the conditional information, Eq. (4.6); the MI between X_1 and X_1, given X_2, is the same as the conditional SMI of X_1, given X_2.

With the two identities (4.5) and (4.6), we can generalize Eq. (4.4) for three rv's as follows:

First, we rewrite the MI in Eq. (4.4) in terms of the two pieces of self-information as shown in Eqs. (4.5) and (4.6), i.e.

$$I(X_1;X_2) = I(X_1;X_1) - I(X_1;X_1|X_2). \tag{4.7}$$

For three rv's, we generalize Eq. (4.7) as

$$CI(X_1;X_2;X_3) = I(X_1;X_2) - I(X_1;X_2|X_3). \tag{4.8}$$

We will refer to the *CI* defined in Eq. (4.8) as the *conditional* MI between the three rv's. The meaning of $CI(X_1;X_2;X_3)$ is a measure of the effect of knowing X_3 on the MI between X_1 and X_2. The definition in Eq. (4.8) is symmetric with respect to X_1 and X_2 but not with respect to X_3. We now show that the conditional MI is in fact symmetric with respect to the three variables. To show this, we use the definitions of the two MI's on the right hand side of Eq. (4.8):

$$\begin{aligned} CI(X_1;X_2;X_3) &= \sum_{x_1,x_2} p(x_1,x_2)\log\left[\frac{p(x_1,x_2)}{p(x_1)p(x_2)}\right] \\ &\quad -\sum_{x_3} p(x_3)\sum_{x_1,x_2} p(x_1;x_2|x_3) \\ &\quad \times\log\left[\frac{p(x_1,x_2|x_3)}{p(x_1|x_3)p(x_2|x_3)}\right]. \end{aligned} \tag{4.9}$$

Using the definitions of the conditional probabilities

$$p(x_1|x_3) = \frac{p(x_1,x_3)}{p(x_3)}, \tag{4.10}$$

$$p(x_2|x_3) = \frac{p(x_2,x_3)}{p(x_3)}, \tag{4.11}$$

$$p(x_1,x_2|x_3) = \frac{p(x_1,x_2,x_3)}{p(x_3)}. \tag{4.12}$$

We can rewrite Eq. (4.9) as

$$\begin{aligned} CI(X_1;X_2;X_3) &= -\sum p(x_1,x_2,x_3) \\ &\quad \times \log\left[\frac{p(x_1,x_2,x_3)p(x_1)p(x_2)p(x_3)}{p(x_1,x_2)p(x_1,x_3)p(x_2,x_3)}\right]. \end{aligned} \tag{4.13}$$

In this form, we see that $CI(X_1;X_2;X_3)$ is symmetric with respect to the three rv's X_1, X_2, and X_3. We can also rewrite it as

$$\begin{aligned} CI(X_1;X_2;X_3) &= I(X_1;X_2) - I(X_1;X_2|X_3) \\ &= I(X_1;X_3) - I(X_1;X_3|X_2) \\ &= I(X_2;X_3) - I(X_2;X_3|X_1). \end{aligned} \tag{4.14}$$

Thus, we can say that $CI(X_1;X_2;X_3)$ measures the difference between the MI between any two rv's, and the conditional MI between the same two variables given the knowledge about the third rv.

4.3 The Relationship between the Conditional MI and the SMI

We next express $CI(X_1;X_2;X_3)$ in terms of SMI. We use Eq. (4.13) to rewrite the CI as

$$\begin{aligned} CI(X_1;X_2;X_3) = &-\sum p(x_1,x_2,x_3)\log p(x_1,x_2,x_3) \\ &-\sum p(x_1)\log p(x_1) - \sum p(x_2)\log p(x_2) \\ &-\sum p(x_3)\log p(x_3) + \sum p(x_1,x_2)\log p(x_1,x_2) \\ &+\sum p(x_1,x_3)\log p(x_1,x_3) \\ &+\sum p(x_2,x_3)\log p(x_2,x_3) \\ = &\, H(X_1) + H(X_2) + H(X_3) - H(X_1,X_2) \\ &- H(X_1,X_3) - H(X_2,X_3) + H(X_1,X_2,X_3). \end{aligned} \quad (4.15)$$

The last form on the right hand side of Eq. (4.15) is reminiscent of the inclusion–exclusion principle in probability. In fact, this form of the CI has motivated Matsuda (2000) and others to define this conditional MI by Eq. (4.15). Also, one is tempted to use the Venn diagram for SMI similarly to its usage in probability. As noted in Chapter 0, the Venn diagram is useful for studying the probabilities of events. It is not recommended to use it for the SMI and the MI, which measure the extent of *dependence* between events.

In the case of probability, if each of the pairs of events A, B and C are mutually exclusive, i.e. disjoint in pairs, namely $A \cap B = A \cap C = B \cap C = 0$, then it follows that the three events A, B, and C are also disjoint, i.e. $A \cap B \cap C = 0$ (the inverse of this statement is not true; three events may have no point in common, but pairs of the three events are not necessarily disjoint). Independence between events behaves differently. If the events are independent in pairs, it does not follow that they are independent in triplets. Also, independence in triplets does not imply independence in pairs [see example in Ben-Naim (2015b)]. Therefore, it is not advisable to use the Venn diagram for the SMI.

As we will see in Sec. 4.8, there are cases where the three rv's are dependent, and yet the triplet MI as defined in Eq. (4.9) is *negative*. Presenting such a case by a Venn diagram would require regions of overlapping between SMI's having a negative area.

4.4 The Connection between the *TI* and the *CI*

The general relationship between the *TI* and the *CI* for the case of three rv's can be obtained from Eq. (4.15):

$$\begin{aligned} CI(X_1;X_2;X_3) = &-\sum p(x_1)\log p(x_1) - \sum p(x_2)\log p(x_2) \\ &-\sum p(x_3)\log p(x_3) + \sum p(x_1,x_2)\log p(x_1,x_2) \\ &+\sum p(x_1,x_3)\log p(x_1,x_3) \\ &+\sum p(x_2,x_3)\log p(x_2,x_3) \\ &-\sum p(x_1,x_2,x_3)\log p(x_1,x_2,x_3). \end{aligned} \quad (4.16)$$

Note also that each singlet probability is the marginal probability of the pair probability, such as

$$p(x_1) = \sum_{x_2} p(x_1,x_2) = \sum_{x_2,x_3} p(x_1,x_2,x_3). \quad (4.17)$$

Therefore, we can rewrite Eq. (4.16) as

$$\begin{aligned} CI(X_1;X_2;X_3) = &\, TI(X_1;X_2) + TI(X_1;X_3) + TI(X_2;X_3) \\ &- TI(X_1;X_2;X_3). \end{aligned} \quad (4.18)$$

Note that the *TI* for two rv's is the same as the MI $I(X_1;X_2)$.

Thus, we have related the *CI* to the *TI* defined with respect to the total correlation functions.

4.5 The Connection between the Conditional MI and the Kirkwood Superposition Approximation

The reader who is not familiar with the Kirkwood superposition approximation may skip this section.

The relationship (4.16) can be transformed into a useful connection between the *CI* and the so-called Kirkwood superposition approximation (KSA). Using the definitions of the pair and triplet correlation functions, we can write Eq. (4.16) as

$$\begin{aligned} CI(X_1;X_2;X_3) &= -\sum p(x_1,x_2,x_3)\log[g(x_1,x_2,x_3)] \\ &\quad +\sum p(x_1,x_2,x_3)\log[g(x_1,x_2)g(x_1,x_3)g(x_2,x_3)] \\ &= -\sum p(x_1,x_2,x_3)\log\left[\frac{g(x_1,x_2,x_3)}{g(x_1,x_2)g(x_1,x_3)g(x_2,x_3)}\right]. \end{aligned} \tag{4.19}$$

The KSA assumes that the triplet correlation $g(x_1,x_2,x_3)$ is a product of the three pair correlations $g(x_1,x_2)g(x_1,x_3)g(x_2,x_3)$, i.e. it assumes that

$$g(x_1,x_2,x_3) = g(x_1,x_2)g(x_1,x_3)g(x_2,x_3). \tag{4.20}$$

Hence, the quantity *CI* in Eq. (4.19) is interpreted as the *average deviation* from the KSA, formulated in terms of the correlation function. It should be noted that in the theory of liquids one defines the pair and the triplet correlation functions as follows:

$$g(\mathbf{R}_1,\mathbf{R}_2) = \frac{p(\mathbf{R}_1,\mathbf{R}_2)d\mathbf{R}_1,d\mathbf{R}_2}{p(\mathbf{R}_1)p(\mathbf{R}_2)d\mathbf{R}_1,d\mathbf{R}_2} \tag{4.21}$$

$$g(\mathbf{R}_1,\mathbf{R}_2,\mathbf{R}_3) = \frac{p(\mathbf{R}_1,\mathbf{R}_2,\mathbf{R}_3)d\mathbf{R}_1,d\mathbf{R}_2,d\mathbf{R}_3}{p(\mathbf{R}_1)p(\mathbf{R}_2)p(\mathbf{R}_3)d\mathbf{R}_1,d\mathbf{R}_2,d\mathbf{R}_3}. \tag{4.22}$$

In the numerators of Eqs. (4.21) and (4.22), we have the probabilities of finding the two or the three particles at the locations $\mathbf{R}_1, \mathbf{R}_2$ and $\mathbf{R}_1, \mathbf{R}_2, \mathbf{R}_3$, respectively. [Note that the probabilities depend on whether we refer to a *specific* particle or *any* particle. However, the ratios in Eqs. (4.21) and (4.22) are the same whether we refer to a *specific* or *any* particle. For details, see Ben-Naim (1992, 2006a).]

Another definition of the correlation functions in Eqs. (4.21) and (4.22) is for particles which are not spherical; say, each particle is described by its locational vector $\mathbf{R}_i$ and its orientation Ω_i (a specific example of a particle having spins will be discussed in Sec. 4.9).

For simplicity, suppose that each particle can have only a finite number of orientations (say, spin up and down). In this case, we define the correlation functions

$$g(\mathbf{R}_1, \mathbf{\Omega}_1, \mathbf{R}_2, \mathbf{\Omega}_2) = \frac{p(\mathbf{R}_1, \mathbf{\Omega}_1, \mathbf{R}_2, \mathbf{\Omega}_2)}{p(\mathbf{R}_1, \mathbf{\Omega}_1)p(\mathbf{R}_2, \mathbf{\Omega}_2)}, \tag{4.23}$$

$$g(\mathbf{R}_1, \mathbf{\Omega}_1, \mathbf{R}_2, \mathbf{\Omega}_2, \mathbf{R}_3, \mathbf{\Omega}_3) = \frac{p(\mathbf{R}_1, \mathbf{\Omega}_1, \mathbf{R}_2, \mathbf{\Omega}_2, \mathbf{R}_3, \mathbf{\Omega}_3)}{p(\mathbf{R}_1, \mathbf{\Omega}_1)p(\mathbf{R}_2, \mathbf{\Omega}_2)p(\mathbf{R}_3, \mathbf{\Omega}_3)}. \tag{4.24}$$

With these correlation functions, we define the CI as a *partial* average, i.e.

$$CI(\mathbf{X}_1; \mathbf{X}_2; \mathbf{X}_3) = -\sum_{\mathbf{\Omega}_1, \mathbf{\Omega}_2, \mathbf{\Omega}_3} p(\mathbf{\Omega}_1, \mathbf{\Omega}_2, \mathbf{\Omega}_3) \log\left[\frac{g(1,2,3)}{g(1,2)g(1,3)g(2,3)}\right]. \tag{4.25}$$

Here, we have used shorthand notations $(1, 2, 3)$ to stand for a specific configuration of the three particles (both locations and orientations). The summation is over all the *orientations* of the three particles. The resulting quantity, denoted as $CI(X_1; X_2; X_3)$, is the average of the deviation from the KSA over all possible *orientations*; the locations of the particles are fixed even after the performance of the averaging process. We will discuss a simple case of three spins having two orientations only (up and down, in Sec. 4.9).

It should be noted that the KSA was used extensively in the theory of liquids. However, there exists no theoretical argument in favor of this approximation. Perhaps Kirkwood (1935) had in mind an approximation for the so-called potential of mean force (PMF). See Ben-Naim (1992, 2006a).

For spherical particles, Eqs. (4.26) and (4.27) define the PMF between two and three particles:

$$W(\mathbf{R}_1,\mathbf{R}_2) = -k_B T \log\left[g(\mathbf{R}_1,\mathbf{R}_2)\right], \tag{4.26}$$

$$W(\mathbf{R}_1,\mathbf{R}_2,\mathbf{R}_3) = -k_B T \log\left[g(\mathbf{R}_1,\mathbf{R}_2,\mathbf{R}_3)\right]. \tag{4.27}$$

It can be shown [see Ben-Naim (1992, 2006a)] that W is related to the work associated with the process of bringing two or three particles from infinite separation to the final configuration $(\mathbf{R}_1,\mathbf{R}_2)$ or $(\mathbf{R}_1,\mathbf{R}_2,\mathbf{R}_3)$, respectively. k_B is the Boltzmann constant and T is the absolute temperature. We leave the base of the logarithm unspecified, but when we interpret W in thermodynamics we use the natural logarithm in Eqs. (4.26) and (4.27). We also note that the work referred to above is usually identified with the Gibbs energy change for the process at constant temperature, pressure, and composition.

One can also show that the gradient of W with respect to, say, $\mathbf{R}_1$ is the average force exerted on particle 1 at $\mathbf{R}_1$, averaged over all configurations of the particles, conditioned on the fixed configuration $\mathbf{R}_1,\mathbf{R}_2$ in the case of Eq. (4.26), or $\mathbf{R}_1,\mathbf{R}_2,\mathbf{R}_3$ in the case of Eq. (4.27).

Using Eqs. (4.26) and (4.27), we can define the deviation from the KSA in terms of the PMF as follows. First, we write the triplet PMF as

$$W(\mathbf{R}_1,\mathbf{R}_2,\mathbf{R}_3) = W(\mathbf{R}_1,\mathbf{R}_2) + W(\mathbf{R}_1,\mathbf{R}_3) + W(\mathbf{R}_2,\mathbf{R}_3). \tag{4.28}$$

This is equivalent to Eq. (4.20), using Eq. (4.27).

This is similar to the pairwise additive assumption for the *interaction energies*, which is considered to be a good approximation. However, it is not a good approximation for the PMF. [See Ben-Naim (2013, 2016b).] We can define the deviation from the KSA by the quantity δW:

$$\delta W(\mathbf{R}_1, \mathbf{R}_2, \mathbf{R}_3) = W(\mathbf{R}_1, \mathbf{R}_2, \mathbf{R}_3) - W(\mathbf{R}_1, \mathbf{R}_2) - W(\mathbf{R}_1, \mathbf{R}_3) - \delta W(\mathbf{R}_2, \mathbf{R}_3). \quad (4.29)$$

Here, δW measures the extent of *deviation* from the pairwise additivity of the triplet PMF. We also have the relationship between the quantity δW and the deviation from the KSA in terms of correlation functions. [See Eqs. (4.26)–(4.29)].

$$\delta W(\mathbf{R}_1, \mathbf{R}_2, \mathbf{R}_3) = -k_B T \log \left[\frac{g(\mathbf{R}_1, \mathbf{R}_2, \mathbf{R}_3)}{g(\mathbf{R}_1, \mathbf{R}_2) g(\mathbf{R}_1, \mathbf{R}_3) g(\mathbf{R}_2, \mathbf{R}_3)} \right]. \quad (4.30)$$

Thus, the quantity CI as defined in Eq. (4.25) for particles having both locations $\mathbf{R}_i$ and orientations Ω_1 may be written as

$$CI(\mathbf{X}_1; \mathbf{X}_2; \mathbf{X}_3) = \sum_{\Omega_1, \Omega_2, \Omega_3} \rho(\Omega_1, \Omega_2, \Omega_3) \beta \delta W(1, 2, 3) = \langle \beta \delta W \rangle. \quad (4.31)$$

where $\beta = (k_B T)^{-1}$.

Note again that $(1, 2, 3)$ stands for the *full description* of the locations and orientations of the three particles. The average in Eq. (4.31) is only an average over all the possible orientations of the three particles.

Thus, in Eq. (4.31), we see that CI measures the average (over all orientations) of the *deviations* from pairwise additivity of the triplet PMF.

4.6 Interpretation of the *CI* in Terms of the MI between Two rv's and the MI between One rv and the Joint rv

We use again the *CI* defined in Eq. (4.13), which we can rewrite as

$$CI(X_1;X_2;X_3) = -\sum p(x_1,x_2,x_3)\log\left[\frac{p(x_1,x_2,x_3)}{p(x_1)p(x_2,x_3)}\frac{p(x_1)p(x_2)}{p(x_1,x_2)}\right.$$
$$\left.\times\frac{p(x_1)p(x_3)}{p(x_1,x_3)}\right]$$
$$= \sum p(x_1,x_2,x_3)\log\left[\frac{p(x_1,x_2)}{p(x_1)p(x_2)}\right]$$
$$+\sum p(x_1,x_2,x_3)\log\left[\frac{p(x_1,x_3)}{p(x_1)p(x_3)}\right]$$
$$-\sum p(x_1,x_2,x_3)\log\left[\frac{p(x_1,x_2,x_3)}{p(x_1)p(x_2,x_3)}\right]. \quad (4.32)$$

Using the definition of the MI between the two rv's we rewrite Eq. (4.32) as

$$CI(X_1;X_2;X_3) = I(X_1;X_2) + I(X_1;X_3) - I(X_1;(X_2,X_3)). \quad (4.33)$$

The last quantity on the right hand side of Eq. (4.33) is defined by

$$I(X_1;(X_2,X_3)) = \sum p(x_1,x_2,x_3)\log\left[\frac{p(x_1,x_2,x_3)}{p(x_1)p(x_2,x_3)}\right]. \quad (4.34)$$

In Eq. (4.33), the quantity *CI* is interpreted as the *difference* between the sum of the MI's $I(X_1;X_2) + I(X_1;X_2)$, and the MI, $I(X_1;(X_2,X_3))$.

Note that on the right hand side of Eq. (4.33) we have only the MI between *two* rv's, i.e. $I(X_1;(X_2,X_3))$ is the amount of information that we gain about X_1 from the knowledge of both X_2 and X_3. This interpretation is useful in applications when one transmits information from two (or more) sources.

4.7 Generalization to n rv's

The generalization of the *TI* for any number of rv's is given in Eq. (4.3). The generalization of the *CI* for any number of rv's starts from the definition in Eq. (4.14):

$$CI(X_1;X_2;X_3) = I(X_1;X_2) - I(X_1;X_2|X_3). \tag{4.35}$$

We next define the *CI* for four rv's as

$$CI(X_1;X_2;X_3;X_4) = CI(X_1;X_2;X_3) - CI(X_1;X_2;X_3|X_4), \tag{4.36}$$

and for any n as:

$$CI(X_1;X_2;\ldots;\mathbf{X}_n) = CI(X_1;X_2;\ldots;\mathbf{X}_{n-1}) - CI(X_1;X_2;\ldots;X_{n-1}|X_n). \tag{4.37}$$

This can be expanded in terms of the SMI of a single rv, or a pair, triplet, etc., of rv's:

$$CI(X_1;X_2;\ldots;X_n) = \sum_{k=1}^{n}(-1)^{k-1} \sum_{(i_1,\ldots,i_k)} H(X_{i_1},X_{i_2},\ldots,X_{i_k}). \tag{4.38}$$

The second sum on the on the right hand side of Eq. (4.38) is over all possible set of indices $(i_1,i_2,\ldots,i_k)\in(1,\ldots,n)$.

For the case $n = 3$, Eq. (4.38) reduces to Eq. (4.15). We note that this expression is reminiscent of the inclusion–exclusion principle in probability. This is the reason some authors use the Venn diagram for the conditional MI.

4.8 Properties of the Multivariate MI

The total MI defined in Eq. (4.3) is always positive for any number of rv's. This follows from the general inequality we had in Sec. 3.3. If *all* the rv's are independent, then the corresponding correlation function $g(x_1, x_2, \ldots, x_n)$ is equal to 1 for all values of $(x_1, \ldots, x_n)$. Therefore, the TI in this case is zero.

Recall that the MI between two rv's is always positive [see Eq. (3.50)]. This is not always true for the CI. For instance, for three rv's [see Eq. (4.33)], if X_1 and X_2 are independent, then $I(X_1; X_2) = 0$. Also when X_1 and X_3 are independent, we have $I(X_1; X_3) = 0$. However, the quantity $I(X_1; X_2, X_3)$ does not have to be zero, even when $I(X_1; X_2) = I(X_1; X_3) = 0$. In such a case, $CI(X_1; X_2; X_3)$ will be *negative*. This is a somewhat counterintuitive result. We expect that if X_1 and X_2 are independent [i.e. $I(X_1; X_2) = 0$], and X_1 and X_3 are also independent [i.e. $I(X_1; X_3) = 0$], then X_1 will also be independent of (X_2, X_3). To put it differently, if knowing X_1 does not convey any information on X_2, and knowing X_1 does not convey any information on X_3, then we expect that knowing X_1 will not convey any information on the joint rv (X_2, X_3); or, equivalently, we expect that knowing (X_2, X_3) will not convey any information on X_1.

Another way of understanding the origin of the negative value of the CI is to examine Eq. (4.18), where on the right-hand side, we have only the total correlation MI. In terms of correlation functions, it is clear that if each of the pair correlations is 1, it does not follow that the triplet correlation will also be 1. Hence, from Eq. (4.18), we get again a negative CI. If there are no correlations between each pair in Eq. (4.18), then we "expect" that there should be no correlation between the triplet, and Eq. (4.18) is negative. This case is sometimes referred to as "frustration" [Matsuda (2000)]. We will see a few qualitative examples of frustrated systems in Sec. 4.9. Here, we have discussed only two qualitative examples. It should be said that in the literature the conditional MI appears under different names, such as "*Interaction of information*" [McGill (1954)], or "*coinformation*" [Bell (2003)]. Sometimes, the terms "*redundancy*" or "*synergy*" are used for this quality.

Note also that in this book we use the definition of $CI(X_1; X_2; X_3)$ as in Eq. (4.14). Some authors define the quantity $(-CI(X_1; X_2; X_3))$ which is referred to as the *interaction information* [McGill (1954)].

A positive $CI(X; Y; Z)$ occurs when knowing Z diminishes the average correlation between X and Y, i.e.

$$I(X; Y) > I(X; Y|Z). \tag{4.39}$$

Hence, $CI(X; Y; Z)$ defined in Eq. (4.8) is positive. Here are a few qualitative examples that appear frequently in the literature:

(a) The first example:
X: Occurrence of rain.
Y: Occurrence of darkness.
Z: Presence of clouds.

Clearly, X and Y are independent events, and hence $I(X : Y) = 0$. On the other hand, knowing that it is cloudy "hints at" both "dark" and "rainy." Thus, knowing Z (cloudy) induces correlation between X and Y. Hence, $I(X; Y|Z) > 0$, and $CI(X; Y; Z)$ is negative.

(b) The second example:
X: The grass is wet.
Y: The water sprinkler is revolving.
Z: It is raining.

Clearly, $(X; Y) > 0$, i.e. there is some correlation between X and Y.

Also, $I(X; Y|Z) < I(X; Y)$, knowing that Z diminishes the correlation between X and Y.

Hence, we conclude that

$$I(X; Y) - I(X; Y|Z) > 0 \quad CI(X; Y; Z) > 0.$$

(c) The third example:
  X: The car engine fails to start.
  Y: Dead battery.
  Z: Blocked fuel pump.

Usually, Y and Z are independent, i.e. $I(Y;Z) = 0$. Knowing Y provides no information on Z, and vice versa. However, knowing X induces dependence between Y and Z. If we know that the battery is good, it implies that there is a fuel block. If we know that there is not a fuel block, then we know that the battery must be dead. In this case, $I(Y;Z|X) \neq 0$. Hence, $CI(X;Y;Z) = I(Y;Z) - I(Y;Z|X) < 0$. This is a case of *negative CI*.

These examples are very qualitative. It is not clear what the probability distributions are, and it is not clear that the MI in these cases is symmetric with respect to X, Y, and Z. In the next section, we will study a more quantitative example of three rv's for which we can calculate all the relevant MI.

4.9 A Three-Spin System

In this section, we will study a *quantitative* example of three rv's for which we know all the probability distributions. Hence, we can calculate all the relevant SMIs, as well as the relevant MI.

In the first part of this section, we present the method of calculating all the relevant probabilities. This is based on statistical thermodynamics. The reader does not need to understand the method we use to calculate these probabilities. He or she is urged only to check that the calculated probabilities *make sense*. Once these probabilities are accepted, the reader is urged to carefully examine all the derived quantities, and do some of the suggested exercises. I hope that this study will be rewarding to the reader in understanding all the quantities defined in this chapter.

The system is an extension of the two-spin system we discussed in Sec. 3.4. Here, we have three spins situated at the vertices of a regular triangle as shown in Fig. 4.1.

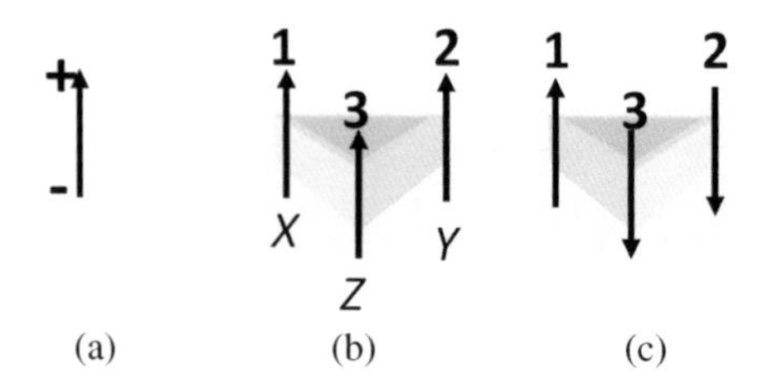

Fig. 4.1. A system of three spins: (a) The dipole, (b) high energy state, and (c) low energy state.

The total number of configurations is $2^3 = 8$; see Fig. 4.2. Each spin can be in either the "up," or the "down" state. Assigning the value of $+1$ to the "up" state, and of -1 to the "down" state, we write the *interaction energy* for any configuration of the three spins as

$$U(x, y, z) = -J \times (x \times y + x \times z + y \times z). \tag{4.40}$$

As we did in Sec. 3.4, $J > 0$ corresponds to a ferromagnet and $J < 0$ to an antiferromagnet. Note that we do not have any external field in this case. However, in the study of this system, we will not be interested in the physical meaning of the system, but only in the probabilities of different events and the corresponding SMI and the MI. All the probabilities are derived from the following equation:

$$p(x, y, z) = \frac{\exp[-\beta U(x, y, z)]}{Z_3}, \tag{4.41}$$

where Z_3 is the normalization constant, defined as

$$Z_3 = \sum_x \sum_y \sum_z \exp[-\beta U(x, y, z)]. \tag{4.42}$$

As in Sec. 3.4, we take $\beta = \frac{1}{T}$, where T is the absolute temperature, and we put the Boltzmann constant $k_B = 1$. Compare Eqs. (4.41) and (4.42) with the corresponding equations (3.57) and (3.58). The

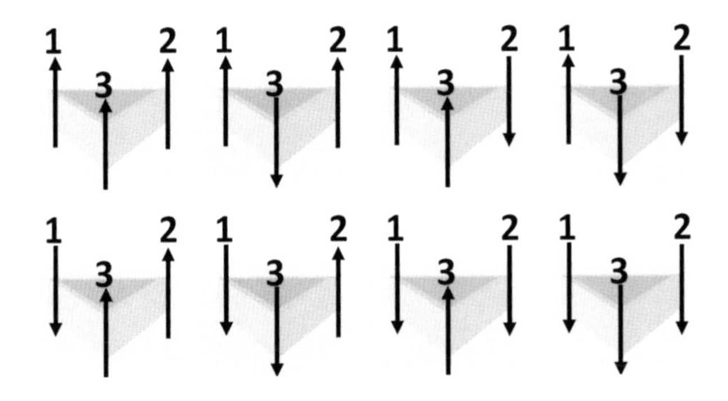

Fig. 4.2. All eight configurations of the three spins.

sum in Eq. (4.42) is over all possible values of x, y, z which in our case are ± 1. Thus, $p(x, y, z)$ is the probability of finding the spin number 1 (X) to be in the state x ("up" or "down"), the spin number 2 (Y) to be in the state y, and the spin number 3 (Z) to be in state z. Hence,

$$p(x, y, z) = P(X = x, Y = y, Z = z). \tag{4.43}$$

Note that all the eight configurations in Fig. 4.2 are of two types; either all three spins are in the same state ["up–up" and "down–down" in Fig. 4.1(b)], with interaction energy $-3J$, or one spin is in one state ("up" or "down"), and the other two are in the other state ("down" or "up"); see Fig. 4.1(c). There are six such configurations in Fig. 4.2. The corresponding energy of the latter configuration is $+J$. Thus, Z_3 may be written as

$$Z_3 = 2\exp(+3\beta J) + 6\exp(-3\beta J). \tag{4.44}$$

For $T \to 0$ and $J > 0$, the two configurations of the type in Fig. 4.1(b) have the *higher* probability. For $T \to 0$ and $J < 0$, the six configurations of the type in Fig. 4.1(c) have the *higher* probability. Such configurations are often referred to as being *frustrated*, in the sense that not all *pairs* of spins can be "satisfied," i.e. be in opposite directions.

4.9.1 *Probabilities*

All the triplet probabilities are calculated from Eq. (4.41). Once we know $p(x, y, z)$, we can calculate all the marginal probabilities. The pair probability is defined as

$$p(x, y) = \sum_{z} p(x, y, z). \tag{4.45}$$

Here, the sum is over all possible values of z. Similarly, $p(y)$ and $p(z)$ are defined as the marginal probabilities of $p(x, y)$.

Figure 4.3 shows the singlet probabilities $p(X = 1)$ and $p(X = -1)$, which we abbreviate to $p(1)$ and $p(-1)$. As in the case of two spins discussed in Sec. 3.4, the value of $p(1)$ [and $p(-1)$] is 0.5 independently of the temperature (T), and of the strength of the interaction (J), and it is the same for X, Y, and Z.

Figure 4.4 shows the pair probabilities as a function of T for different values of positive J. In these plots, we show the pair probabilities, say $p(1, 1)$ or $p(1, -1)$, without specifying which pair of spins we are referring to. Because of the symmetry of the triplet of spins, the same probabilities apply to any pair: (X, Y), (X, Z), or (Y, Z). When $J = 0$ (no interactions), all four configurations are equally likely, and hence $p(1, 1) = p(1, -1) = p(-1, 1) = p(-1, -1) = 0.25$. This is the same as in the case of two spins discussed in Sec. 3.4.

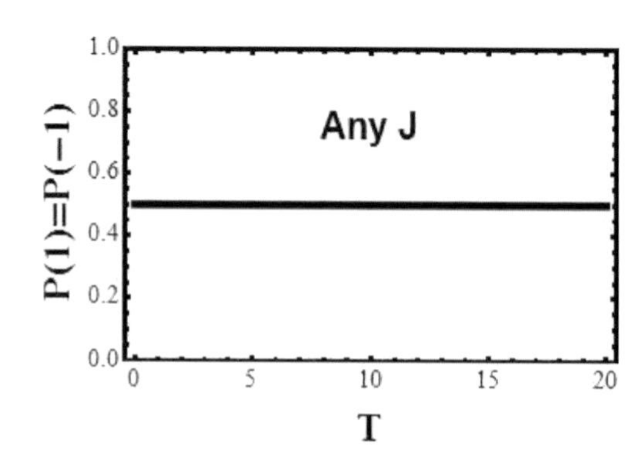

Fig. 4.3. The singlet distribution as a function of T.

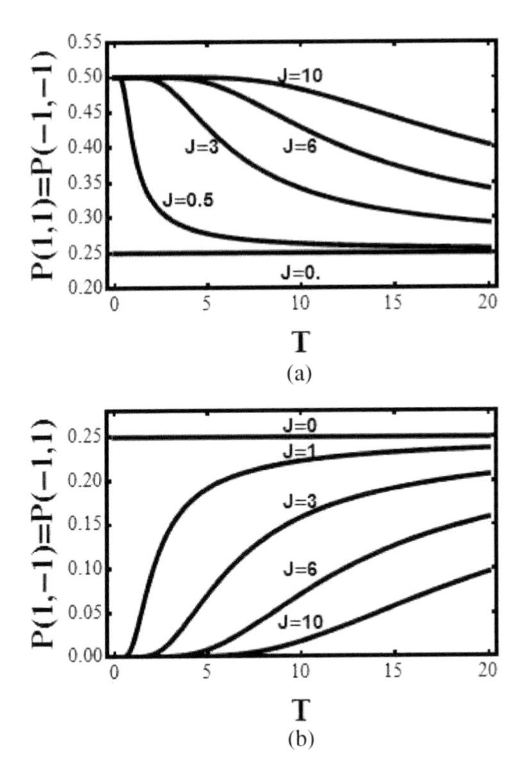

Fig. 4.4. The pair distribution as a function of T for different positive values of J.

For any $J > 0$, all the curves start at a value of 0.5 in the cases "up–up" and "down–down," but at value 0 for the cases "up–down" and "down–up." This is plausible since $J > 0$ favors the same orientation of the two spins and disfavors the opposite orientations. In all cases, we see that as $T \to \infty$ all the curves converge to 0.25, i.e. all four possible configurations of the two spins become equally likely.

Figure 4.5 shows similar values of the pair probabilities for the case $J < 0$. Here, the behavior of the pair probabilities is different from the case of two spins (see Sec. 3.4). As expected, for $J = 0$, all the four configurations are equally probable. Turning on the interaction parameter $J < 0$ favors the antiparallel configurations.

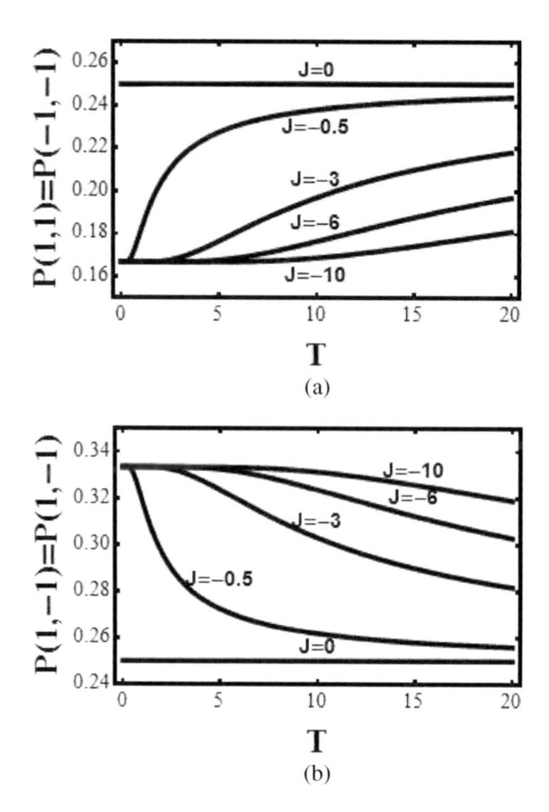

Fig. 4.5. The pair distribution as a function of T for different negative values of J.

However, unlike the case of two spins, here the "turning" on of $J < 0$ has a different effect than we observed in the two-spin system.

For the "up–up" and "down–down" configurations, the probabilities are *lower* than the initial values of 0.25, but unlike the case of two spins (see Sec. 3.4), here the lowest values of the pair probabilities are not zero but $1/6 \approx 0.166$. The reason is that at low temperatures the *all-parallel* configurations are eliminated, and we are left with six configurations of equal probabilities (1/6). These are shown in Fig. 4.6.

For the antiparallel configurations, "up–down" and "down–up" we see again that for $J = 0$, these configurations have equal

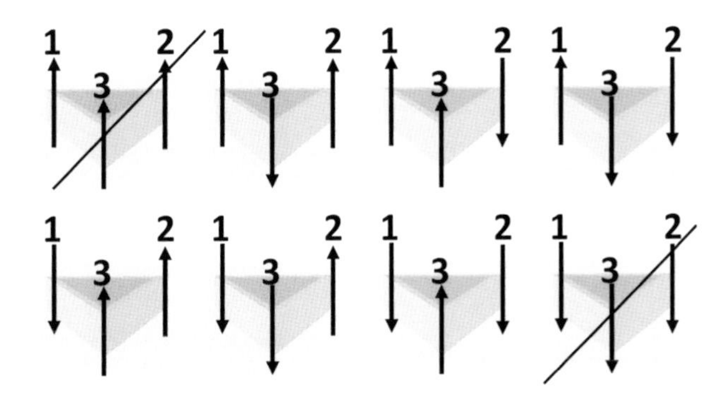

Fig. 4.6. All eight possibilities are equally probable when $J = 0$, any T. For $J = -10$, $T = 0$, two configurations are impossible, (the first and the last), leaving only six configurations with equal probability.

probabilities of 0.25. However, turning on the interactions ($J < 0$) increases the probabilities, not to 0.5 as in the case of two spins [see Fig. 3.8(b)], but to the limiting value of $1/3 = 0.333$. The reason is that at very low temperatures the all-parallel configurations (Fig. 4.6) are eliminated, and we are left with six possible configurations. Therefore, when we ask about $p(1, -1)$ there are only two out of six possible configurations, i.e. probability $2/6 = 1/3$. As we increase the temperature, all configurations tend to be equally probable. The reader should examine these results before proceeding to study the triplet probabilities.

Next, we turn to the triplet probabilities. Figure 4.7(a) shows the probability of "all-up" (or "all-down") configurations. As before, when $J = 0$ the probability is $1/8 = 0.125$, simply because we have eight equally probable configurations. Turning on the interactions ($J > 0$), we see that at low temperatures ($T \to 0$) the probability is 0.5. The reason is that in this case only two configurations ("all-up" and "all-down") are possible, having equal probabilities. As the temperature increases, all the curves in Fig. 4.7(b) tend to the limit of 0.125, i.e. equal probability for all the eight configurations.

Figure 4.7(b) shows the probability $p(-1, 1, 1)$. As expected, for $J = 0$ we have $p(-1, 1, 1) = 0.125$. Turning on the interaction ($J > 0$)

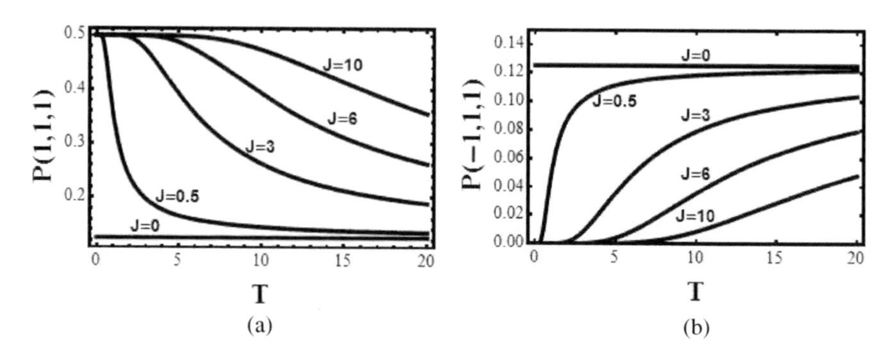

Fig. 4.7. The triplet probabilities: (a) $P(1,1,1)$ and (b) $P(-1,1,1)$, as a function of T for different positive values of J.

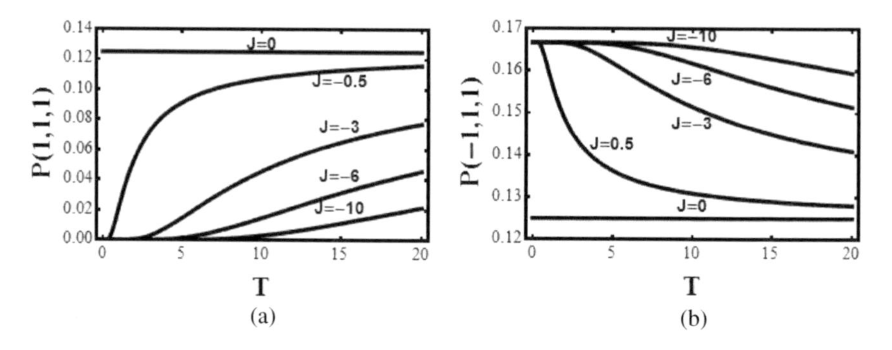

Fig. 4.8. The triplet probabilities: (a) $P(1,1,1)$ and (b) $P(-1,1,1)$ as a function of T for different negative values of J.

makes the (unmarked) six configurations in Fig. 4.6 improbable; only the two marked configurations in Fig. 4.6 are possible. Therefore, we have zero probability for the six unmarked configurations in Fig. 4.6.

Exercise 4.1

Show that the curves for $p(-1,-1,1)$ are the same as for $p(-1,1,1)$ in Fig. 4.7.

Figure 4.8(a) shows $p(1,1,1)$ as a function of T for different values of $J < 0$. In this case, at $T \to 0$ the two marked configurations in Fig. 4.6 have zero probability. The reason is simple. When $J < 0$, the parallel configurations are not favored. The same is true for $p(-1,-1,-1)$. Figure 4.8(b) shows the probability $p(-1,1,1)$ for the case $J < 0$.

When $J < 0$, the parallel configurations are not favored. The same is true for $p(-1,-1,-1)$. Figure 4.8(b) shows the probability $p(-1,1,1)$ for the case $J < 0$. Again, when $J = 0$, the probabilities of all eight configurations are the same and equal to 0.125. However, when we turn on the interaction we see that at low temperatures ($T \to 0$) the probability is $1/6 \approx 0.166$. The reason is that only six unmarked configurations in Fig. 4.6 are possible, and they are equally probable.

Exercise 4.2

Show that the curves for $p(-1,-1,1)$ should be the same as for $p(-1,1,1)$ in Fig. 4.8(b).

As the temperature increases, there is a randomization effect which eventually leads to equal probabilities for all *eight* configurations in Fig. 4.6.

4.9.2 *Conditional Probabilities*

The understanding of the conditional probabilities is important for understanding the conditional MI. The reader is urged to study these probabilities very carefully.

Before we examine the conditional probabilities in the three-spin system, it is advisable to look again at the conditional probabilities for the two-spin system (Sec. 3.4).

In the three-spin case, the effect of the presence of a third spin on the conditional probability of say, $p(X = 1|Y = 1)$ is significant. Note that this conditional probability of $X = 1$, given $Y = 1$ in the presence of a third spin. In some cases, we will use a special notation; see below.

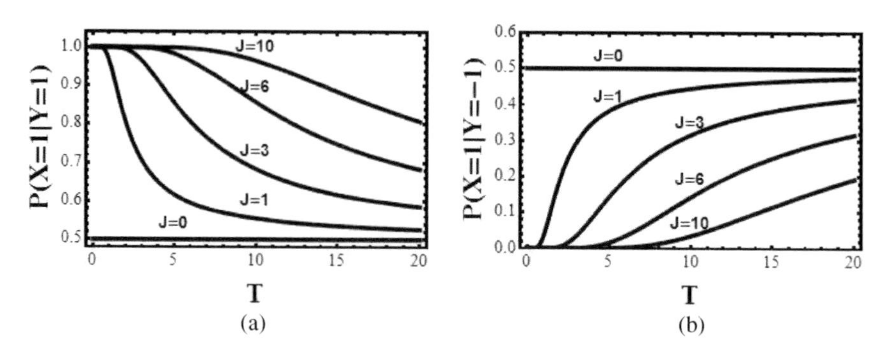

Fig. 4.9. The conditional probabilities: (a) $P(1|1)$ and (b) $P(1|-1)$ as a function of T for different positive values of J.

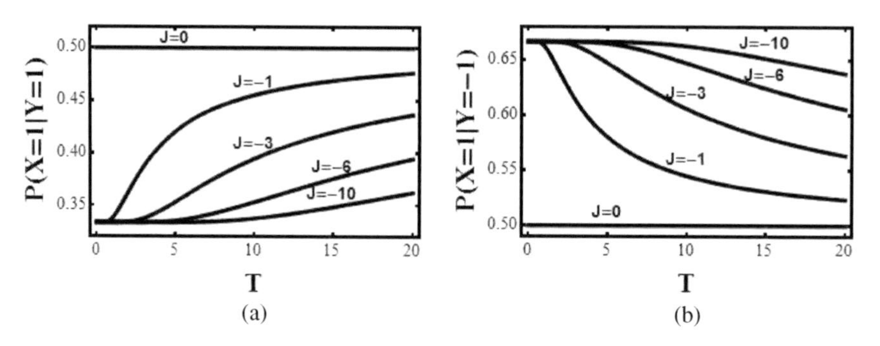

Fig. 4.10. The conditional probabilities: (a) $P(1|1)$ and (b) $P(1|-1)$, as a function of T for different negative values of J.

Figure 4.9 shows $p(X = 1|Y = 1)$ and $p(X = 1|Y = -1)$ as a function of T for various positive values of J. The range of variation is between 0.5 and 1 for the probability $p(X = 1|Y = 1)$, and between 0 and 0.5 for the probability $p(X = 1|Y = -1)$. Thus, at very low temperatures ($T \to 0$), the conditional probability $p(X = 1|Y = 1)$ is 1, independently of the presence of a third spin. Likewise, the conditional probability $p(X = 1|Y = -1)$ for $T \to 0$ is 0, independently of the presence of a third spin.

The situation is markedly different for the case $J < 0$. Figure 4.10(a) shows the conditional probability $p(X = 1|Y = 1)$ as a function of T for different values of $J < 0$. When there are only two spins, the conditional probability $p(X = 1|Y = 1)$ behaves differently. In this case, the range of values of $p(X = 1|Y = 1)$ is between 0 and 0.5. It is zero at $T \to 0$ because the negative J favors opposite orientations of the two spins. Therefore, if $Y = 1$, the probability of $X = 1$ is zero. When a third spin is present, the situation is different. The upper value at $J = 0$ is 0.5. However, when $J < 0$ and $T \to 0$ the lowest value of $p(X = 1|Y = 1)$ is not zero, but $1/3 = 0.333$. To see why, we examine again Fig. 4.2. Given that $Y = 1$ (spin 2 in Fig. 4.11, in red, is up) eliminates four configurations. Of these four configurations, one has probability 0 (all up) and of the left, only three have their spins up. The second from the left in the first row in Fig. 4.11). These three conformations have the same probability (at $T \to 0$), and one of them is the required event $(X = 1|Y = 1)$.

Figure 4.10(b) shows the conditional probability $p(X = 1|Y = -1) = p(X = -1|Y = 1)$. In this case, when $J < 0$ and $T \to 0$ the upper limit is not 1.0 as in the case of two spins, but $2/3 = 0.666$. The reason can be seen from Fig. 4.11.

Again, there are four possible configurations for which $Y = 1$ (red up in Fig. 4.11). Of these, four are impossible ("all-up"), and out of

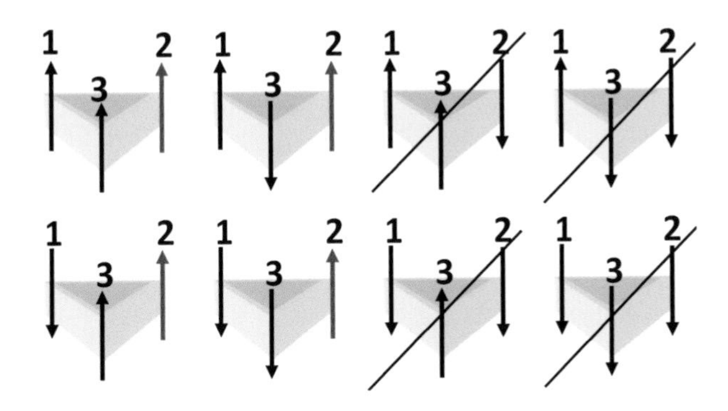

Fig. 4.11. All eight possibilities are equally probable when $J = 0$, any T. For $J = -10$, $P(X = 1|Y = 1)) = 1/3$, but $P(X = -1|Y = 1) = 2/3$. The four marked configurations are impossible.

the three possible ones with $Y = 1$, only two have $X = -1$. Hence, the conditional $p(X = -1|Y = 1)$ in the limit of $T \to 0$ and $J < 0$ is 2/3.

Exercise 4.3

Show that the same argument holds for $p(X = 1|Y = -1)$.

These examples are important for understanding the conditional information and mutual information discussed in the next section. It is also important to understand the differences between the conditional probability of $p(x|y)$ in the two-spin system and in the three-spin system. In the latter case, it is advisable to denote this conditional probability by $p(x|y, \_)$, where the blank space stands for the *presence*, but with an unspecified orientation of the third spin.

Figure 4.12(a) shows the conditional probability $p(X = 1| Y = 1, Z = 1)$. The range of values of this conditional probability is easily understood. Given $Y = 1$ and $Z = 1$, and $J > 0$, and $T \to 0$, we have only two possibilities for X ("up" or "down"); only one of these is possible at $T \to 0$ (the left one in the first row of Fig. 4.11), i.e. $p(X = 1|Y = 1, Z = 1) = 1.0$.

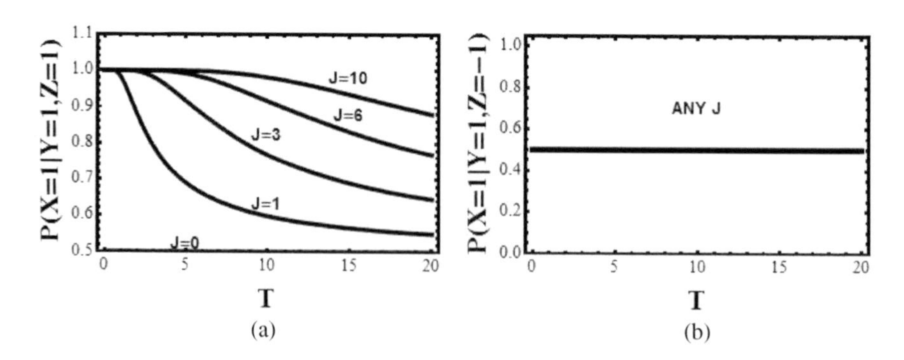

Fig. 4.12. The conditional probabilities: (a) $P(1|1,1)$ and (b) $P(1|1,-1)$ as a function of T for different positive values of J.

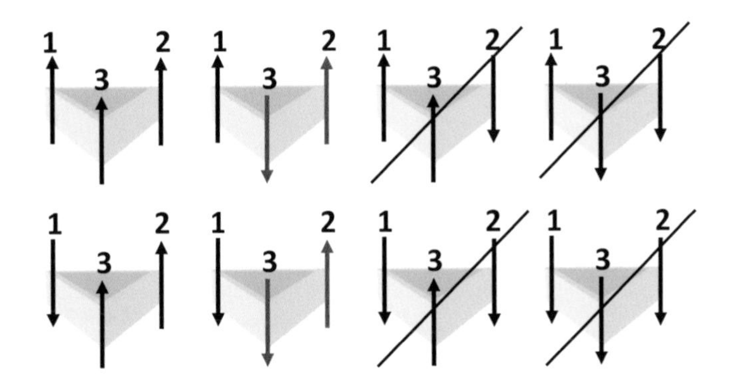

Fig. 4.13.

Exercise 4.4

Show that the same behavior is expected for $p(X = 1|Y = -1, Z = -1)$.

Figure 4.12(b) shows $p(X = 1|Y = 1, Z = -1)$ for any value of J and T. Given that $Y = 1$ and $Z = -1$ (marked in red in Fig. 4.13), there are only two possibilities for X (up or down). When $J > 0$ (any J), we have one favored pair of spins (either up–up or down–down) and two unfavored pairs (antiparallel). When $J < 0$, we have one unfavored interaction (parallel) and two favored pairs (antiparallel), and these have the same probability.

Figure 4.14 shows the conditional probabilities $p(X = 1|Y = 1, Z = 1)$ and $p(X = 1|Y = -1, Z = -1)$. It is left as an exercise for the reader to explain the behavior of these conditional probabilities with the help of Fig. 4.2.

4.9.3 *The SMI and the Conditional SMI*

Based on the probability distributions we have calculated in the previous subsections, we can calculate all the pertinent SMI's. Before studying the SMI in this system, it is helpful to have a glance at the corresponding figures for the two-spin system in Sec. 3.4.

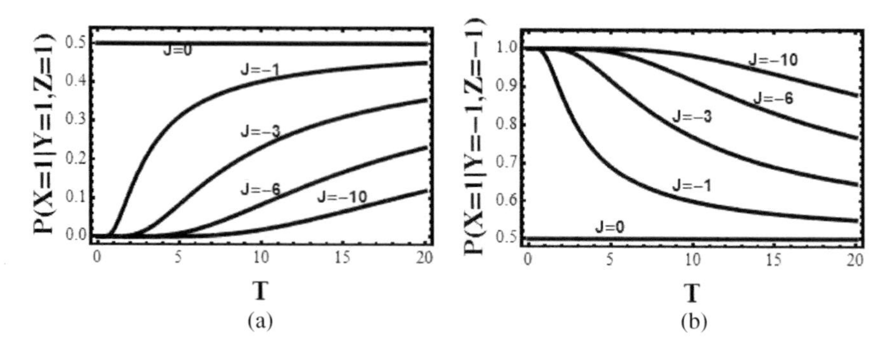

Fig. 4.14. The conditional probabilities: (a) $P(1|1,1)$ and (b) $P(1|-1,-1)$, as a function of T for different negative values of J.

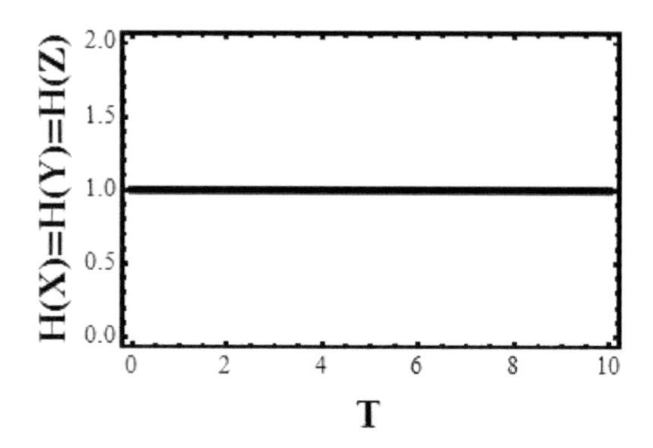

Fig. 4.15. The SMI for one spin as a function of T.

We start with the SMI of the single spin. Figure 4.15 shows $H(X) = H(Y) = H(Z)$. As expected, the SMI for a single spin is always 1 (i.e. equal probability for the two states "up" and "down").

Figure 4.16 shows the conditional SMI, $H(X|Y = 1,\_)$. The behavior of this quantity is similar to the case of the two spin system; the bar in the brackets is to remind us that here a third spin is present, but its state is unspecified. For $J > 0$ and $T \to 0$, given $Y = 1$, the most probable configuration is "all-up."

The same is true for $H(X|Y = -1,\_)$. Therefore, the average of the two, $H(X|Y,\_)$ has the same behavior as $H(X|Y = 1,\_)$. Hence,

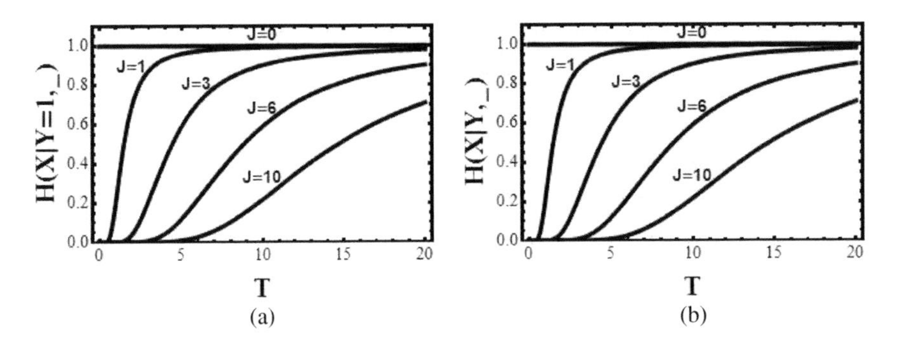

Fig. 4.16. The conditional SMI: (a) $H(X|Y = 1\_)$ and (b) $H(X|Y,\_)$, as a function of T for different positive values of J.

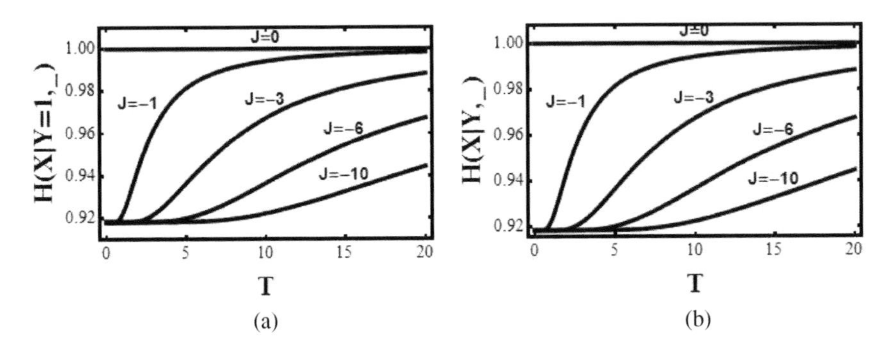

Fig. 4.17. The conditional SMI: (a) $H(X|Y = 1\_)$ and (b) $H(X|Y,\_)$, as a function of T for different negative values of J.

we have the equality

$$H(X|Y = 1,\_) = H(X|Y = -1,\_) = H(X|Y,\_). \tag{4.46}$$

A different behavior is observed for $J < 0$. The relevant data are shown in Fig. 4.17. Here, the region of variation of the conditional SMI, $H(X|Y = 1,\_)$ is between ≈ 0.92 and 1. To understand this behavior, we examine all the four configurations for which $Y = 1$ (spin 2 is up); see Fig. 4.18. For $J < 0$ and $T \to 0$, the "all-up" configuration is unstable, i.e. it has probability 0. From the remaining three configurations which are equally probable, the conditional

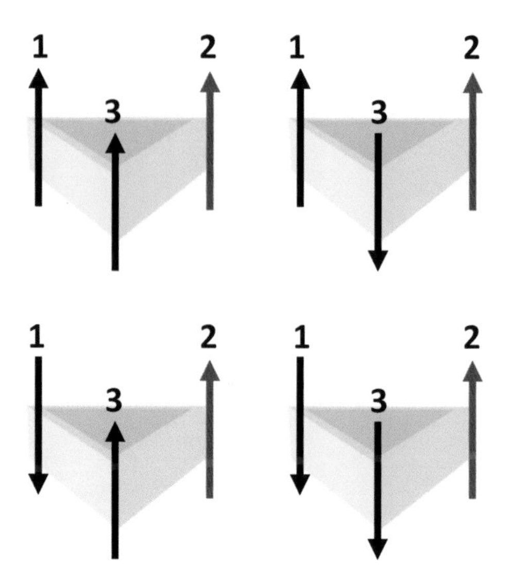

Fig. 4.18.

probability of $p(1|1,\_)$ is $1/3$ (only the right configuration in the first row in Fig. 4.18), and $p(-1|1,\_) = 2/3$ (the two configurations in the second row in Fig. 4.18). Therefore, the conditional SMI at $T \to 0$ and $J < 0$ is

$$H(X|Y = 1,\_) = -\frac{1}{3}\log\frac{1}{3} - \frac{2}{3}\log\frac{2}{3} \approx 0.92. \tag{4.47}$$

Note also that the same equality as in Eq. (4.46) holds for this case.

Next, we calculate the conditional SMI, $H(X|Y = 1, Z = 1)$; see Fig. 4.19(a).

At $T \to 0$ and $J > 0$, we have only one stable configuration ("all-up"), and therefore the values of the SMI start at $H(X|Y = 1, Z = 1) = 0$. On the other hand, the SMI $H(X|Y = 1,\ Z = -1)$ or $H(X|Y = -1,\ Z = 1)$ is equal to 1, see Fig. 4.19(b). In this case, when $(Y = 1, Z = -1)$, there are only two equally likely probabilities; see Fig. 4.19(c).

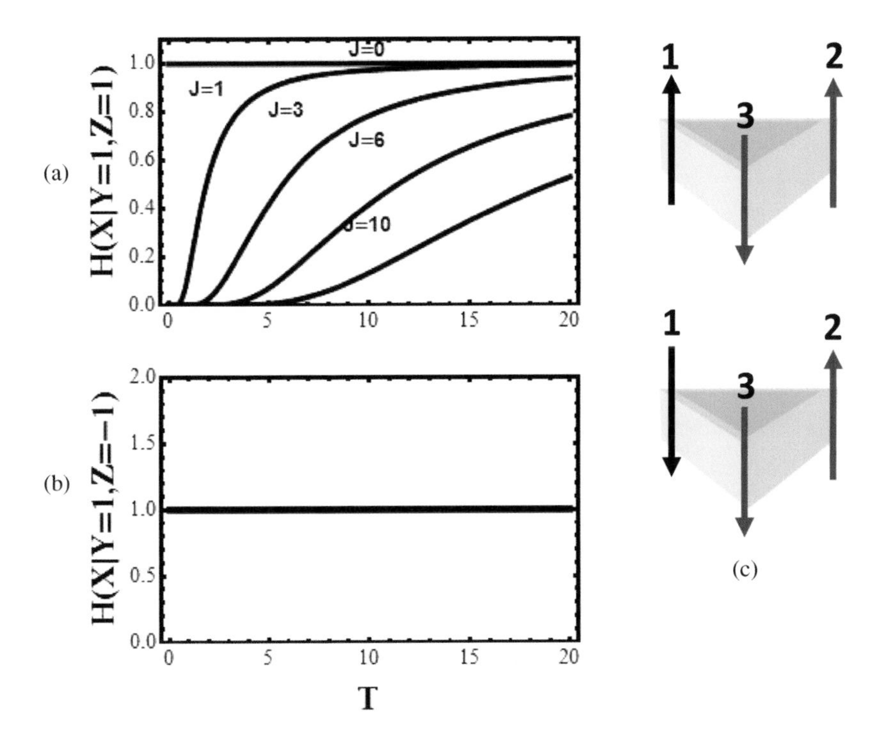

Fig. 4.19. The conditional SMI: (a) $H(X|Y = 1, Z = 1)$; (b) $H(X|Y, Z = -1)$, as a function of T for different positive values of J; and (c) the two configurations having equal probabilities.

Exercise 4.5

Examine the behavior of the conditional SMI, $H(X|Y = x, Z = z)$, for the case $J < 0$ and for all possible pairs of Y and Z : $(1, 1)$ $(-1, -1), (-1, 1)$ and $(1, -1)$.

Next, we turn to the SMI of two spins in the presence of the third one.

Figure 4.20(a) shows $H(X, Y|Z = 1)$ for $J \geq 0$.

The range of variation of the values of this SMI is between 0 and 2. Given $Z = 1$, $J > 0$, and $T \to 0$, we have only one configuration with probability 1; this is the "all-up" configuration and the corresponding value of the SMI is 0. For higher temperatures we have

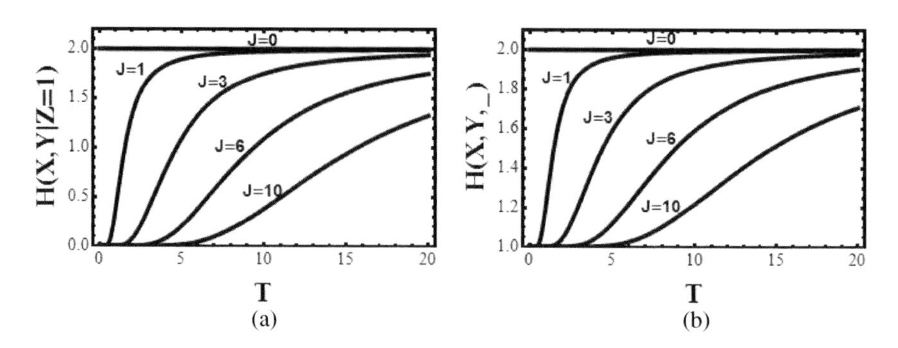

Fig. 4.20. The conditional SMI: (a) $H(X, Y|Z = 1)$, and (b) $H(X, Y, \_)$, as a function of T for different positive values of J.

equal probabilities for all the four configurations of X and Y given $Z = 1$.

It is easy to see that the same curves are obtained for $H(X, Y|Z = -1)$, and therefore also for the average of those cases. Hence, we have the equality

$$H(X, Y|Z = 1) = H(X, Y|Z = -1) = H(X, Y|Z). \tag{4.48}$$

Another conditional SMI which should be considered is shown in Fig. 4.20(b). This is denoted by $H(X, Y, \_)$, which means the joint SMI of X and Y given the *presence*, but unspecified state, of Z. This should be compared with the joint SMI shown in Fig. 3.6 for a two-spin system. Note the difference between the range of values of $H(X, Y|Z$ and of $H(X, Y, \_)$. At $T \to 0$, *knowing* Z leaves only one stable configuration, (either X and Y "up" in the case $Z = 1$, or X and Y "down" in the case $Z = -1$). On the other hand, *knowing* only the *presence* of the third spin, we have at $T \to 0$ two possible stable configurations, *both* X and Y "up" and X and Y "down," and therefore in this case we have $H(X, Y, \_) = 1$ (at $T \to 0$, and $J > 0$).

The reader should note the difference between the meaning of $H(X, Y|Z)$ and on $H(X, Y, \_)$. In the first, we take the average of $H(X, Y|Z = 1)$ and $H(X, Y|Z = -1)$ with the probabilities $p(Z = 1)$ and $p(Z = -1)$. In the second, we calculate the joint SMI for the four

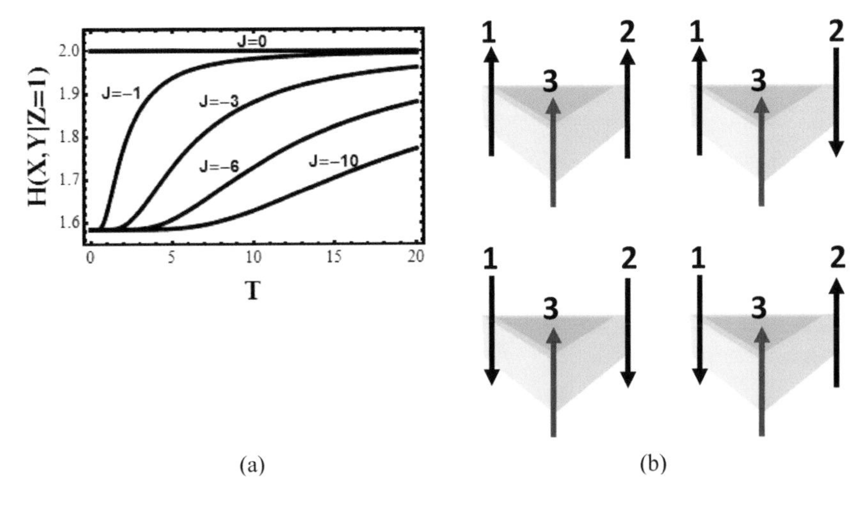

Fig. 4.21. The conditional SMI: (a) $H(X,Y|Z=1)$ as a function of T for different negative values of J, and (b) the relevant configurations.

possible states $(X=1;Y=1)(X=1;Y=-1)$, $(X=-1;Y=1)$, and $(X=-1;Y=-1)$ with the pair probabilities, $p(x,y)$, and we form the average

$$H(X,Y,\_) = -\sum\sum p(x,y)\log p(x,y), \tag{4.49}$$

where $p(x,y)$ are the marginal probabilities computed in the system of three spins, Eq. (4.45).

A very different behavior is observed for $J<0$; see Fig. 4.21(a). Here, given $Z=1$, at $T\to 0$, we have only three stable configurations, [Fig. 4.21(b)] with equal probabilities.

Therefore,

$$H(X,Y|Z=1) = H(X,Y|Z=-1) = H(X,Y|Z) = \log 3 \approx 1.58. \tag{4.50}$$

Note that the conditional SMI, $H(X,Y|Z)$ is the average of the two conditional SMIs on the left hand side of Eq. (4.50). On the other hand, the joint SMI in the *presence* of Z, denoted by $H(X,Y,\_)$ is different, see Figure 4.22(a).

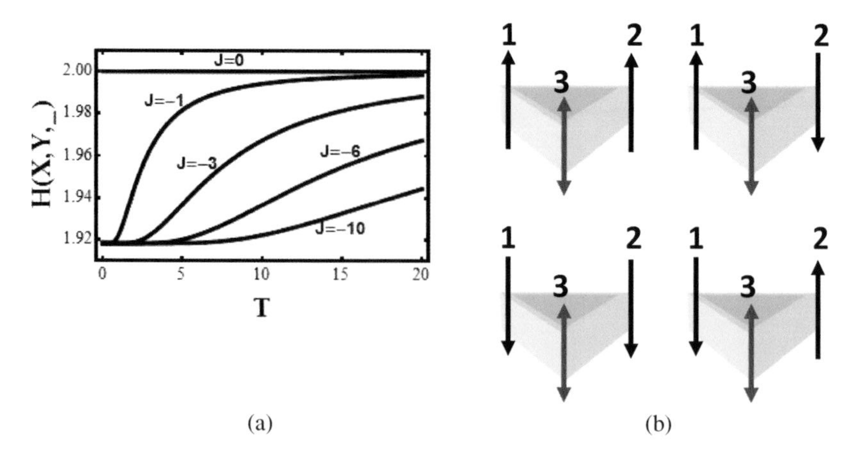

Fig. 4.22. (a) The joint SMI $H(X, Y, \_)$, in the presence of a third spin (the red spin in (b)) as a function of T for different negative values of J.

In Figure 4.22(a), we see that the limiting value of $H(X, Y, \_)$ is about 1.92 (for $J < 0$ and $T \to 0$). The reason is that when Z is present we have four possible states of the two spins X and Y. These are shown in Fig. 4.22(b). Note the double-headed arrow of spin 3. This indicates that the spin is present but its state is unspecified. The probability of the parallel configuration is 1/6 and we have two of such configurations [left column of Fig. 4.22(b)]. There are also two antiparallel configurations, each with probability 1/6. Thus, a joint SMI in the *presence* of Z is

$$H(X, Y, \_) = -2\left(\frac{1}{6}\log\frac{1}{6} + \frac{1}{3}\log\frac{1}{3}\right) \approx 1.92. \qquad (4.51)$$

The reader should examine the behavior of the different SMI's in Figs. 4.21 and 4.22.

Finally, in Fig. 4.23, we show the joint SMI for the three spins. For $J > 0$ and $T \to 0$, there are only two stable states with equal probabilities: "all–up" and "all–down." Hence, in this limit,

$$H(X, Y, Z) = \log 2 = 1 \text{ (for } J > 0 \text{ and } T \to 0). \qquad (4.52)$$

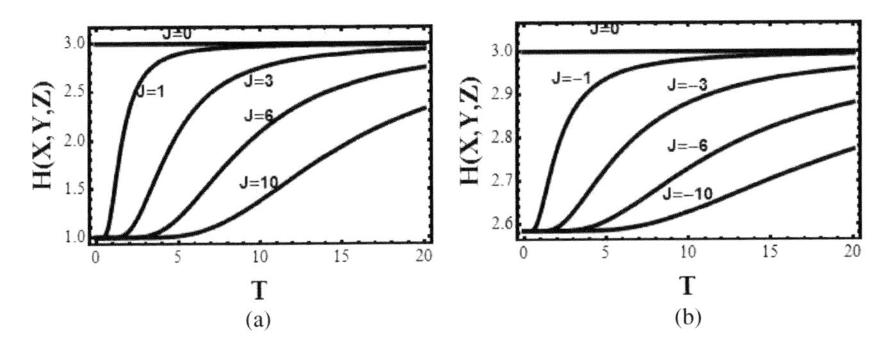

Fig. 4.23. The joint SMI for the three spins $H(X,Y,Z)$ as a function of T for different (a) positive values, and (b) negative values of J.

As the temperature increases, the value of $H(X,Y,Z)$ tends to 3, i.e. all eight possibilities become equally probable.

A different behavior is shown for $J < 0$. Here, at $T \to 0$, the two configurations "all-up" and "all-down" are excluded, leaving six configurations having equal probabilities. Hence, the limiting value in this case is

$$H(X,Y,Z) \approx 2.58 \quad (J < 0;\ T \to 0). \tag{4.53}$$

As the temperature increases, we see again that the values of $H(X,Y,Z)$ increase to the limiting value of 3.

4.9.4 *Correlations and the* MI

Before we calculate the various MI's in this system, it is advisable to examine the pair correlations in the three-spin system.

Figure 4.24 shows some data on the correlations $g(x,y)$ in the three-spin system; see Fig. 3.8. Note first that the pair correlations in Fig. 4.24 were calculated in the presence of a third spin. The general behavior of pair correlations in Fig. 4.24 is similar to that in Fig. 3.8, but not identical. For instance, the values of the correlations $g(1,1)$ for $J > 0$ range between 1 and 2, which is similar to the behavior in Fig. 3.8. On the other hand, the range for $g(1,-1)$ is between 0.66 and 1 in Fig. 4.24, compared to between 0 and 1 in Fig. 3.8. Thus, the

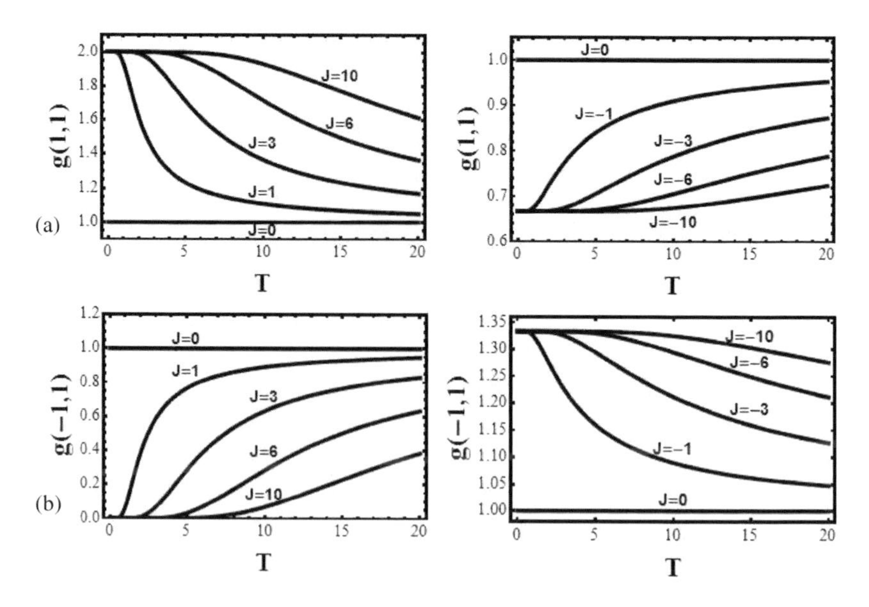

Fig. 4.24. The various pair correlation function as a function of T for different (a) positive, and (b) negative values of J.

presence of a third spin has an effect on the correlation between the two spins X and Y.

Similarly, the range of variation of $g(-1,1)$ is between 0 and 1 ($J > 0$) for both the two-spin and the three-spin system. On the other hand, the range of values of $g(-1,1)$ for the case $J < 0$ is different. For the two-spin system, the range is between 1 and 2, whereas for the three-spin system it is between 1 and 1.333.

Figure 4.25 shows the MI of X and Y in the presence of a third spin (but in an unspecified state). This should be compared with $I^{(2)}(X;Y)$ in Fig. 3.9. The latter notation ($I^{(2)}$) is used to remind us that this MI is calculated in a two-spin system.

Likewise, we could have written $I^{(3)}(X;Y)$ for the MI between X and Y calculated in the three-spin system. However, in order to avoid confusion with the MI for the three spins (see below), we use the notation $I(X;Y,\_)$ where the bar stands for the presence of a third spin, the state of which is unspecified.

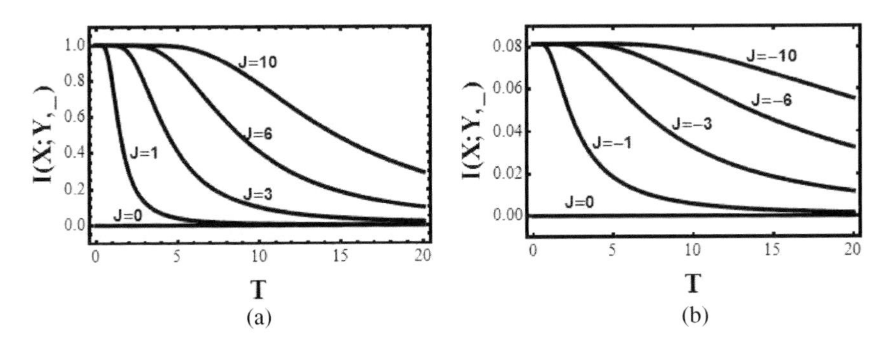

Fig. 4.25. The mutual information in the presence of a third spin as a function of T for different (a) positive, and (b) negative values of J.

In Fig. 4.25(a), we show the MI as a function of T for $J > 0$. We see that the general behavior of $I(X;Y,\_)$ is similar to that of $I^{(2)}(X;Y)$ in Fig. 3.9. When $J = 0$, there is no correlation between the two spins, and hence $I(X;Y,\_) = 0$. On the other hand, turning on the interaction ($J > 0$) induces a correlation between X and Y. When $T \to 0$, the MI is 1, which means that knowing X provides full information on Y, and vice versa. Unlike the case of two-spin systems (Fig. 3.9), where we observed the same behavior for $J > 0$ and $J < 0$, here, in the presence of a third spin, the behavior of $I(X;Y,\_)$ is very different in the case $J < 0$; see Fig. 2.25(b).

First, note that in the range of values in Fig. 4.25(b), we see that when we have strong interaction $J = -10$, the maximum MI is 0.08. This means that strong interaction ($J \approx -10$) does not induce a large MI; knowing X provides little information on Y, and vice versa.

Exercise 4.6

Calculate $I(X;Y,\_)$ for a system of three spins for which there is interaction only between X and Y, but no interaction between X and Z, or Y and Z. The potential function in this system is

$$U(x,y,z) = -J(x \times y). \tag{4.54}$$

This exercise is important. Turning off the interactions between X and Z, and between Y and Z is equivalent to the *absence* of a third spin. The corresponding results for the MI are the same as in Chap. 3, i.e. for the two-spin system.

***Exercise* 4.7**

Calculate $I(X; Y, \_)$ as well as other MI's for the case where there is *no* interaction between X and Y, but there are interactions between X and Z, and between Y and Z. The potential function in this case is

$$U(x, y, z) = -J(x \times z + y \times z). \tag{4.55}$$

This exercise is also important. Here, we can learn that although there is no *direct* interaction between X and Y, there could be correlation, and hence MI between X and Y, which is due to the presence of the third spin, Z.

The reader should pause and think about the difference in behavior of the MI in Figs. 4.25(a) and 4.25(b). One way of understanding the different behavior is to do the two exercises given above, where we *separated* the effect of the *direct* interaction from that of the indirect interaction. Now, consider the case of very low temperature, $T \to 0$. When $J > 0$, knowing $X = 1$ induces two effects on Y, and the direct interaction between X and Y will cause Y to favor "up." In addition, X will also cause Z to prefer "up," and Z in turn will cause Y to favor "up." Thus, we see that both the direct and the indirect interactions contribute to *positive* correlation between X and Y.

Next, consider the case $X = -1$ ("down"). This causes Y to prefer "down." It also causes Z to favor "down," which in turn causes Y to be "down." We see again that both the direct and the indirect interaction produce positive correlation between X and Y. This is the reason for the very large value of $I(X; Y, \_)$ in Fig. 2.25(a) ($J > 0$ and $T \to 0$).

Now, let us examine the case $J < 0$ and $T \to 0$ knowing that $X = 1$ ("up") causes Y to prefer "down." In addition, $X = 1$ causes

Z to favor "down," which pushes Y toward "up." We see that the direct and the indirect interaction induce opposite effects. The direct interaction induces negative correlation, whereas the indirect interaction induces positive correlation.

Next, still with $J < 1$, suppose that we know that $X = -1$ ("down"). The direct interaction causes Y to prefer "up" (negative correlation). In addition, $X = -1$ causes Z to favor "up," and therefore Z pushes Y toward "down." Hence, the direct interaction produces positive correlation.

Thus, in the presence of Z, we see that knowing X provides much information on Y when $J > 0$, but very little information when $J < 0$.

Next, we examine the conditional MI. Figure 4.26 shows the behavior of $I(X;Y|Z = 1)$ and $I(X;Y|Z)$ for $J > 0$. In this case, we have the equality

$$I(X;Y|Z = 1) = I(X;Y|Z = -1) = I(X;Y|Z). \tag{4.56}$$

As can be seen, the effect of knowing Z on the MI between X and Y is very small. To understand this, we use the definition

$$\begin{aligned} I(X;Y|Z) &= H(X|Z) + H(Y|Z) - H(X,Y|Z) \\ &= 2H(X|Z) - H(X,Y|Z). \end{aligned} \tag{4.57}$$

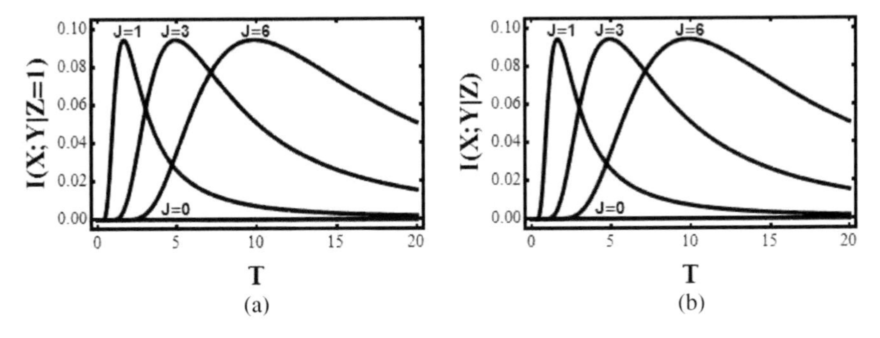

Fig. 4.26. The conditional mutual information as a function of T for different positive values of J.

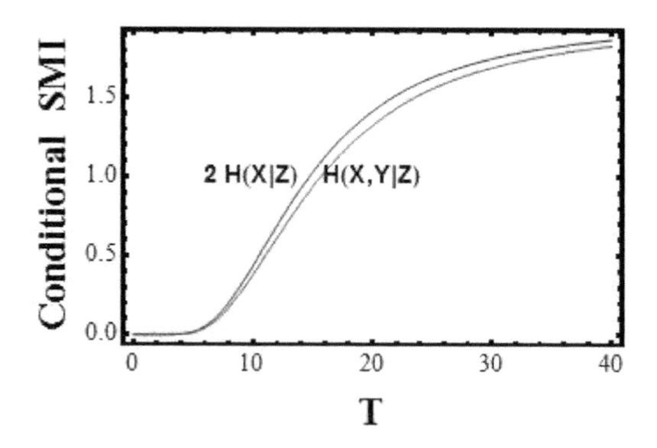

Fig. 4.27. The conditional SMI as a function of T.

Figure 4.27 shows the values of $2H(X|Z)$ (blue line) and $H(X, Y|Z)$ (red line) for the choice of $J = 10$.

We see that both curves increase with temperature but at slightly different rates. We can also see that the difference between $2H(X|Z)$ and $H(X, Y|Z)$ is initially zero, then becomes increasingly positive and at some temperature starts to decrease, and eventually becomes zero at $T \to \infty$.

Note that the range of variation of $(X; Y|Z)$, in Fig. 4.26, is quite small. We can say that the condition Z has a negligible effect on the MI between X and Y.

Figure 4.28 shows similar data for the case $J < 0$. Again, the equality in Eq. (4.56) holds in this case. Unlike the case $J > 0$, we see that here the conditional MI starts at a value of 0.25 at $T \to 0$, then decreases monotonically as the temperature increases. Figure 4.29 shows the conditional SMI $2H(X|Z)$ and $H(X, Y|Z)$, for the case $J < 0$, as a function of T. We see that both of these two quantities increase with temperature. However, the difference between the two is

$$I(X; Y|Z) = 2H(X|Z) - H(X, Y|Z). \tag{4.58}$$

This quantity decreases with temperature, as we are seen in Figs. 4.28(a) and 4.28(b). See also Fig. 3.29.

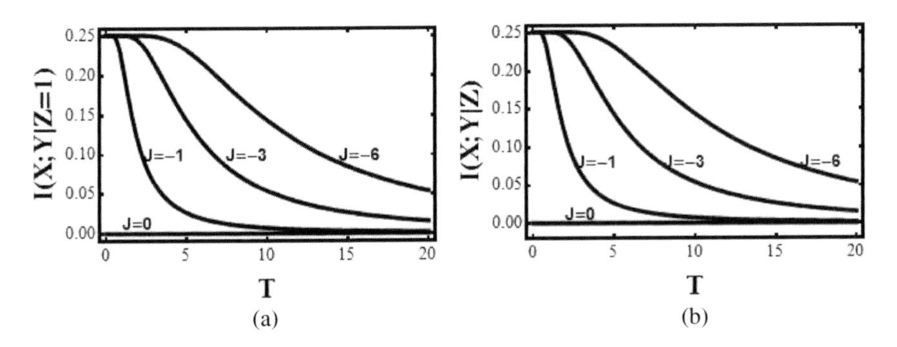

Fig. 4.28. The conditional mutual information as a function of T for different negative values of J.

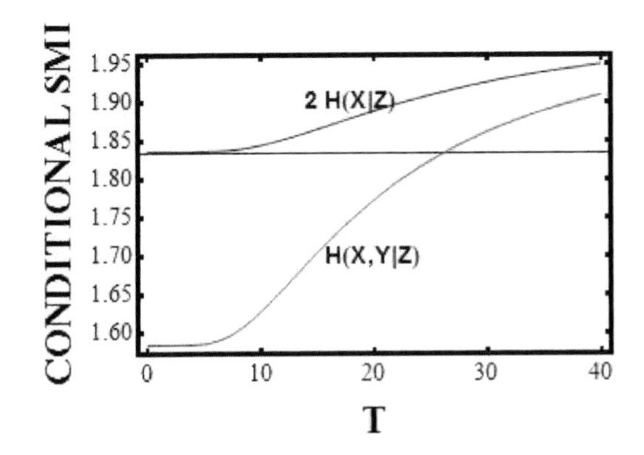

Fig. 4.29. The conditional SMI as a function of T.

Figure 4.30 shows the values of the conditional MI defined by

$$CI(X;Y;Z) = I(X;Y) - I(X;Y|Z). \tag{4.59}$$

For $J > 0$, we saw that $I(X;Y|Z)$ is negligibly small Therefore, CI in Eq. (4.59) is dominated by $I(X;Y)$, which is positive. This is essentially the same as $I(X;Y,\_)$ in Fig. 4.25(a). On the other hand, for $J < 0$, the mutual correlation $I(X;Y)$, which is the same as $I(X;Y,\_)$ in Fig. 4.25(b) is negligibly small. In this case, the quantity $I(X;Y|Z)$ dominates the CI in Eq. (4.59), which is *negative*. The fact that CI is negative is attributed to the *frustration* in the three-spin system.

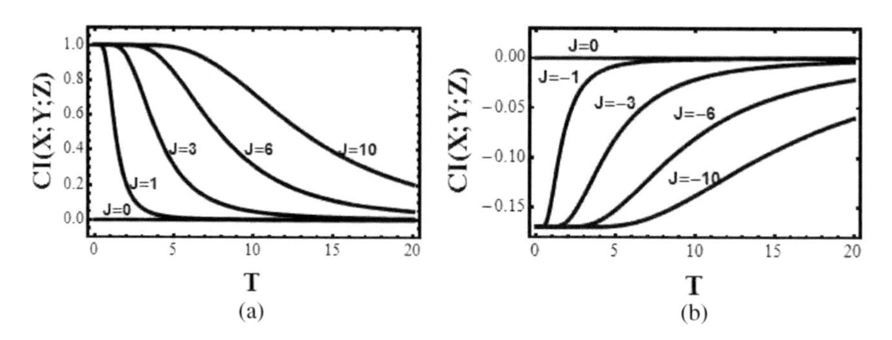

Fig. 4.30. The conditional MI as a function of T for (a) positive and (b) negative values of J.

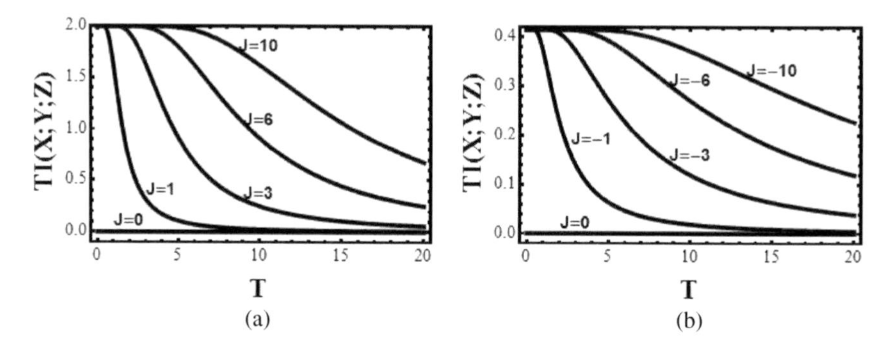

Fig. 4.31. The total MI as a function of T for (a) positive, and (b) negative values of J.

Finally, we show in Fig. 4.31 the total MI between X, Y and Z, defined by

$$TI(X;Y;Z) = \sum_{x,y,z} p(x,y,z) \log[g(x,y,z)]. \tag{4.60}$$

These quantities must always be positive (see Sec. 3.4). We see that this MI is large and positive for $J > 0$, but relatively small for $J < 0$. The reason is that the triplet correlations in the case $J > 0$ are much larger than in the case $J < 0$.

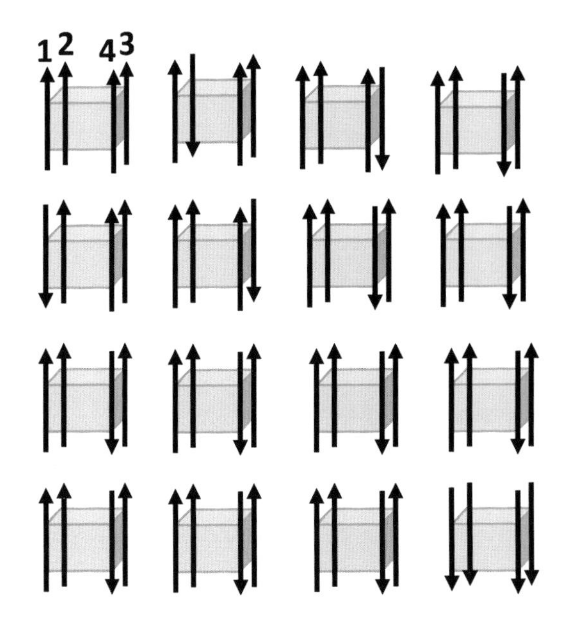

Fig. 4.32. All 16 configurations of the four spins.

4.10 A Four-Spin System

We generalize the system of Sec. 4.9 to four spins situated on the vertices of a square shown in Fig. 4.32. Each spin can be in either the "up" or the "down" state. The total number of configurations is $2^4 = 16$. These are shown in the figure. Again, we assign the value of $+1$ to the "up" state and -1 to the "down" state. The total interaction energy for any configuration (x_1, x_2, x_3, x_4) is given by

$$U(x_1, x_2, x_3, x_4) = -J \times (x_1 \times x_2 + x_2 \times x_3 + x_3 \times x_4 + x_4 \times x_1). \tag{4.61}$$

This is a generalization of Eq. (4.40). Note that we have taken interactions between nearest neighbors only, and J has the same significance as in Eq. (4.40).

All the probabilities for this system can be obtained from the equation

$$p(x_1,x_2,x_3,x_4) = \frac{\exp[-\beta U(x_1,x_2,x_3,x_4)]}{Z_4}, \tag{4.62}$$

with the normalization constant being defined by

$$Z_4 = \sum_{x_1}\sum_{x_2}\sum_{x_3}\sum_{x_4} \exp[-\beta U(x_1,x_2,x_3,x_4)], \tag{4.63}$$

where each sum is over the two values of $x_i = \pm 1$. Substituting Eq. (4.61) in Eqs. (4.62) and (4.63), we see that the products of β and J always appear together. Therefore, in the following figures we will draw all the quantities as a function of βJ (increasing βJ corresponds to either increasing the value of J or decreasing the temperature T).

Unlike the system of three spins, this system does not exhibit the frustration phenomenon. For $\beta J > 0$, the lowest energy configuration is the one where all spins are either "up" or "down." For $\beta J < 0$, the lowest energy is obtained for the antiparallel orientation of each pair of consecutive spins, e.g. "up," and "down," "up," and "down."

The general form of Z_4 is

$$Z_4 = 12 + 2\exp(-\beta J) + 2\exp(+\beta J). \tag{4.64}$$

Thus, there are two configurations with energy J ("all-up" and "all-down") and there are two configurations with energy $+J$ [$(1,-1,1,-1)$ and $(-1,1,-1,1)$], and all the other 12 configurations have zero interaction energy.

From Eq. (4.62), we can obtain all the required probabilities by taking the marginal probabilities of $p(x_1,x_2,x_3,x_4)$. From these probabilities, we can calculate all the relevant SMI's.

Figure 4.33 shows the SMI for one, two, three, and four spins in the four-spin system. Note that all the SMI's for one spin $H(X_1) = 1$, i.e. there is equal probability for the "up" and the "down" states (note that when there is an external field these two probabilities are different, depending on the strength and direction of the field).

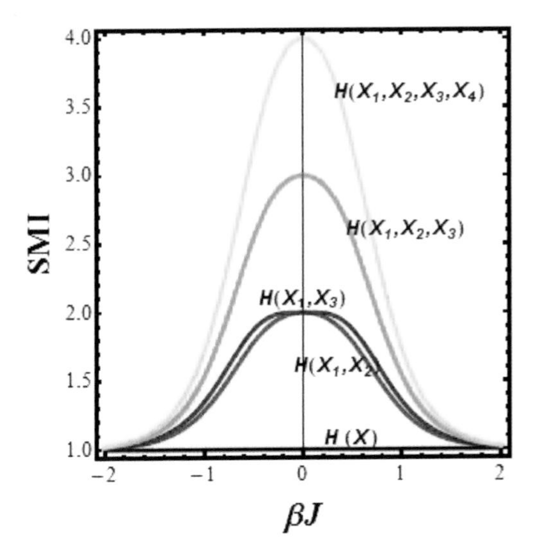

Fig. 4.33. The SMI for one, two, three, and four spins in the four-spin system of Fig. 4.32.

There are two different kinds of the pair SMI: the nearest neighbors, $H(X_I, X_2)$, and the second nearest neighbors, $H(X_I, X_3)$. Both of these attain the value of 1 at $\beta J = 0$ (either $J = 0$ or infinite T) and the value of 0 at $\beta J = \pm\infty$ (either zero temperature or $J = \pm\infty$).

There is only one kind of triplet SMI, say $H(X_I, X_2, X_3)$, which attains the maximum value of 3 at $\beta J = 0$, and one quadruplet SMI, $H(X_I, X_2, X_3, X_4)$, which attains the maximum value of 4 (total randomization of the four spins).

Next, we turn to the various MI's. There are two kinds of pairs of MI's, which we denote by $I^{(4)}(X_I; X_2)$ and $I^{(4)}(X_I; X_3)$. These are the MI's between nearest and second nearest neighbors, respectively. Note that the notation $I^{(4)}$ means that these MI's are calculated in a system of four spins, e.g. the MI between X_1 and X_2 in the presence of X_3 and X_4. These two are shown in Fig. 4.34. We see that two these MI's are almost the same. The values range from 0, at $\beta J = 0$ (independence between two spins), to 1, at a low temperature or high value of J, indicating total dependence.

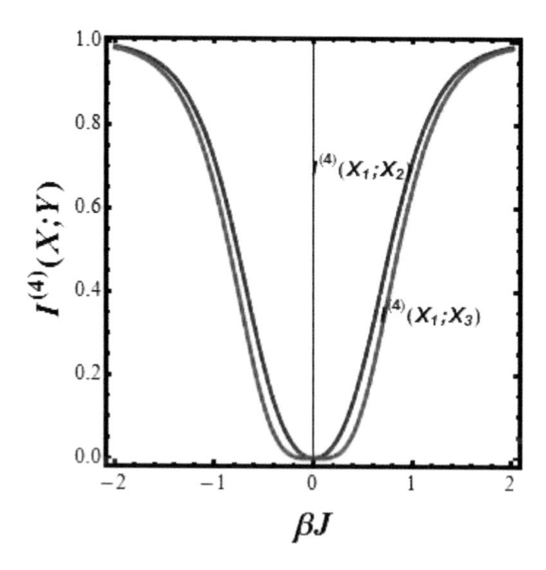

Fig. 4.34. The nearest and next nearest neighbors MI for two spins in the four-spin system.

Figure 4.35 shows the total MI for the triplet and quadruplet spins; note that $TI(X_I;X_2;X_3) = TI(X_I;X_2;X_4)$ have the same values. The curves are similar to the ones in Fig. 4.34, for the pair MI, only the ranges of values are different. Note also that all the values are positive.

In Fig. 4.36, we present the values of the conditional MI for the triplet and quadruplet spins. These are defined by

$$\begin{aligned} CI(X_I;X_2;X_3) = CI(X_I;X_2;X_4) &= 3H(X_I) \\ &- H(X_I,X_2) - H(X_2,X_3) - H(X_1,X_3) \\ &+ H(X_I,X_2,X_3), \end{aligned} \tag{4.65}$$

$$\begin{aligned} CI(X_I;X_2;X_3;X_4) &= 4H(X_I) - 4H(X_I,X_2) \\ &= 2H(X_I,X_3) + 4H(X_I,X_2,X_3) \\ &- H(X_I,X_2,X_3,X_4). \end{aligned} \tag{4.66}$$

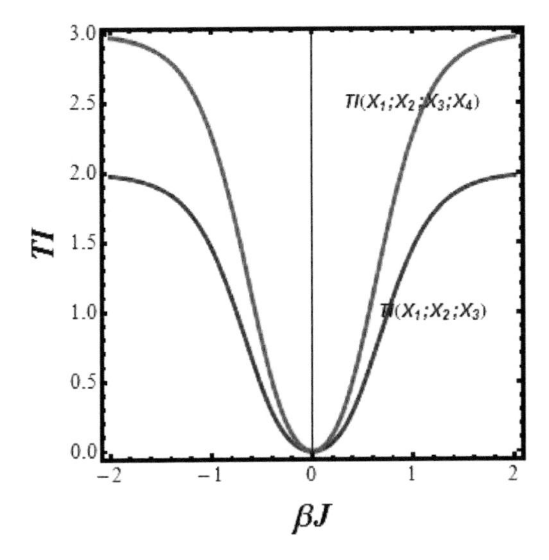

Fig. 4.35. The triplet (blue) and the quadruplet (red) MI in the four-spin system.

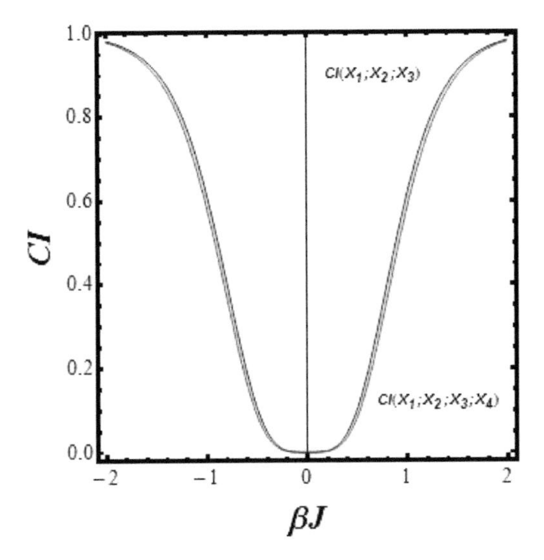

Fig. 4.36. The conditional triplet (blue) and quadruplet (red) MI in the four-spin system.

Note that these are all positive, in contrast to the negative values we got for the three-spin system.

4.11 Summary of Chapter 4

In this chapter, we have not introduced any new concepts. We generalized the quantities we had defined in earlier chapters. These quantities are important in applications of IT in many fields which are not related to communication theory.

5

Entropy and the Second Law of Thermodynamics

In this chapter, we *define* the concept of entropy as a special case of SMI. This definition leads naturally to a simple, intuitive, and irrefutable interpretation of entropy. It also provides a solid probabilistic interpretation of the second law of thermodynamics.

As was explained in Chap. 1, Shannon *called* his measure — which we referred to as Shannon's measure of information (SMI) — "entropy." This was a great mistake, which caused great confusion in both IT and thermodynamics. SMI is defined for any probability distribution. Entropy is defined on a tiny subset of all the possible distributions. Calling SMI "entropy" leads to many awkward statements, such as: "The entropy tends to increase" or "The *value* of the maximum entropy at equilibrium is the *entropy* of the system."

The correct statement concerning the entropy of an isolated system is as follows. An isolated system at equilibrium is characterized by a fixed energy E, volume V, and number of particles N (assuming a one-component system). For such a system, the entropy is *determined* by variables E, V, N (see Sec. 5.2 for a particular case of an explicit entropy function of an ideal gas). This entropy is fixed for that system.

It is not a function of time, it does not change with time, and it does not tend to a maximum. (See Note 1.)

Similarly, one can define the entropy for any other well-defined thermodynamic system at equilibrium. This is exactly what is meant by the statement that entropy is a *state function*.

For any isolated system not at equilibrium, one can define the SMI on the probability distributions of locations and velocities of all the particles. This SMI might change with time. At equilibrium, it attains a maximum value. The maximum value of the SMI attained at equilibrium is related to the entropy of the system.

In this chapter, we use for convenience the natural logarithm $\log_e x$, or $\ln x$. Whenever we want to convert to SMI, we need to multiply by $\log_2 e$, i.e. $\log_2 x = \log_2 e \log_e x$.

5.1 A Few Historical Milestones

There are essentially three definitions of entropy. The first is referred to as the thermodynamic, experimental, macroscopic, or nonatomistic definition. This definition originated in the 19th century, stemming from the interest in heat engines. The introduction of entropy into the vocabulary of physics is due to Clausius. In reality, Clausius did not *define* entropy itself, but only changes in entropy. His definition, together with the third law of thermodynamics, led to the definition of "absolute values" of entropy.

The second definition is attributed to Boltzmann. This definition is sometimes referred to as the microscopic definition or the atomistic definition of entropy. It relates the entropy S of a system to the number of accessible microstates W of a thermodynamic system characterized macroscopically by the total energy E, volume V, and total number of particles N [for a multicomponent system, N may be reinterpreted as the vector $(N_1, \ldots, N_c)$, where N_i is the number of atoms of type i]. Boltzmann's postulated the following relationship, which is now known as Boltzmann's entropy: $S = k_B \log W$, where k_B is a constant, now known as Boltzmann's constant, and W is the number of microstates of the system. Here, log is the natural logarithm.

Boltzmann's definition seems to be completely unrelated to Clausius' definition, yet it is found that for all processes, for which the entropy changes, can be calculated by using Boltzmann's definition, the results agree with the entropy changes, calculated using Clausius' definition. Although there is no formal proof that Boltzmann's entropy is equal to the thermodynamic entropy, it is widely believed that this is true.

The third definition is based on Shannon's measure of information. It may also be referred to as the microscopic or the atomistic definition of entropy. However, this definition of entropy, as well as the formulation of the second law, is very different from Boltzmann's definition. We will often refer to it as the *informational* definition of entropy.

Unlike Boltzmann's definition, the SMI definition does not rely on calculations of the number of accessible microscopic states of the system. It provides directly the *entropy function* of an ideal gas and, by extension, also the entropy function for systems of interacting particles. In this chapter we discuss only the third definition of entropy.

5.2 Derivation of the Entropy Function of an Ideal Gas

The overall plan for obtaining the entropy of an ideal gas from the SMI consists of four steps:

First, we calculate the *locational* SMI, associated with the *equilibrium* distribution of locations of all the particles in the system.

Second, we calculate the *velocity* SMI, associated with the *equilibrium* distribution of velocities (or momenta) of all the particles.

Third, we add a correction term due to the quantum-mechanical *uncertainty principle.*

Fourth, we add a correction term due to the fact that the particles are *indistinguishable.*

Note that in the first two steps we use the maximum SMI method to find out the *equilibrium distribution.* Then we use this equilibrium distribution to *evaluate* the corresponding SMI. In the last two steps, we introduce two corrections to the SMI. Once we combine the results of the four steps, we get, up to a multiplicative constant, the *entropy* of an ideal gas.

5.2.1 *The Locational SMI of a Particle in a 1D Box of Length L*

Suppose that we have a particle confined to a 1D "box" of length L. As we discussed in Sec. 2.10, since there are infinite points at which the particle can be within the interval $(0, L)$, the corresponding locational SMI must be infinite. However, we can define, as Shannon did, the following quantity by analogy with the discrete case:

$$H(X) = -\int f(x)\log f(x)dx. \tag{5.1}$$

This quantity might either converge or diverge, but in any case, in practice, we will use only differences between such quantities. (See Appendix C.) In Sec. 2.11, we calculated the density distribution which maximizes the locational SMI, $H(X)$ in Eq. (5.1), and found that

$$f_{\mathrm{eq}}(x) = \frac{1}{L}. \tag{5.2}$$

The use of the subscript eq (for "equilibrium") will be clarified later. The corresponding SMI, calculated by substituting Eq. (5.2) in Eq. (5.1), is

$$H(\text{locations in 1D}) = \log L. \tag{5.3}$$

We now acknowledge that the location X of the particle cannot be determined with absolute accuracy, i.e. there exists a small interval h_x within which we do not care where the particle is. Therefore, we must correct Eq. (5.3) by subtracting $\log h_x$. Thus, we write, instead of Eq. (5.3),

$$H(X) = \log L - \log h_x. \tag{5.4}$$

We recognize that in Eq. (5.4) we have effectively defined $H(X)$ for the *finite* number of intervals $n = L/h$. (See Fig. 2.12.) Note that when $h_x \to 0$, $H(X)$ diverges to infinity. Here, we do not take the mathematical limit; however, we stop at h_x small enough but not zero. Note also that in writing Eq. (5.4) we do not have to specify the units of length, as long as we use the same units for L and h_x.

5.2.2 *The Velocity SMI of a Particle in a 1D Box of Length L*

In Sec. 2.11, we calculated the probability distribution that maximizes the continuous SMI, subject to two conditions:

$$\int_{-\infty}^{\infty} f(x)dx = 1, \tag{5.5}$$

$$\int_{-\infty}^{\infty} x^2 f(x)dx = \sigma^2 = \text{constant}. \tag{5.6}$$

The result is the normal distribution

$$f_{\text{eq}}(x) = \frac{\exp\left(-x^2/2\sigma^2\right)}{\sqrt{2\pi\sigma^2}}. \tag{5.7}$$

Applying this result to a classical particle having average kinetic energy $\frac{m\langle v_x^2\rangle}{2}$, and using the relationship between the standard deviation σ^2 and the temperature of the system,

$$\sigma^2 = \frac{k_B T}{m}, \tag{5.8}$$

we get the equilibrium velocity distribution of one particle in a 1D system:

$$f_{\mathrm{eq}}(v_x) = \sqrt{\frac{m}{2\pi k_B T}} \exp\left(\frac{-mv_x^2}{2k_B T}\right), \tag{5.9}$$

where k_B is Boltzmann's constant, m the mass of the particle, and T the absolute temperature. The value of the continuous SMI for this probability density is

$$H_{\max}(\text{velocity in 1D}) = \frac{1}{2} \log(2\pi e k_B T/m). \tag{5.10}$$

Similarly, we can write the momentum distribution in 1D, by transforming from $v_x \to p_x = mv_x$, to get

$$f_{\mathrm{eq}}(p_x) = \frac{1}{\sqrt{2\pi m k_B T}} \exp\left(\frac{-p_x^2}{2mk_B T}\right) \tag{5.11}$$

and the corresponding maximum SMI:

$$H_{\max}(\text{momentum in 1D}) = \frac{1}{2} \log(2\pi e m k_B T). \tag{5.12}$$

As we have noted in connection with the locational SMI, the quantities (5.10) and (5.12) were calculated using the definition of the *continuous* SMI. Again, recognizing the fact that there is a limit to the accuracy within which we can determine the velocity (or the momentum) of the particle, we correct the expression in (5.12) by subtracting $\log h_p$, where h_p is a small but finite interval:

$$H_{\max}(\text{momentum in 1D}) = \frac{1}{2} \log(2\pi e m k_B T) - \log h_p. \tag{5.13}$$

Note again that if we choose the units of h_p of momentum as: mass length/time, the same as for $\sqrt{mk_B T}$, then the whole expression under the logarithm will be a pure number.

5.2.3 *Combining the SMI's of the Location and Momentum of One Particle in a 1D System*

In the previous two sections, we derived the expressions for the locational and the momentum SMI of one particle in a 1D system. We now combine the two results. Assuming that the location and momentum (or velocity) of the particles are independent events, we write

$$\begin{aligned} H_{\max}(\text{location and momentum}) &= H_{\max}(\text{location}) \\ &\quad + H_{\max}(\text{momentum}) \\ &= \log\left(\frac{L\sqrt{2\pi emk_B T}}{h_x h_p}\right). \end{aligned} \tag{5.14}$$

Recall that h_x and h_p were chosen to eliminate the divergence of the SMI when the location and momentum were treated as continuous rv's.

In writing Eq. (5.14), we assumed that the location and momentum of the particle are independent. However, quantum mechanics imposes a restriction on the accuracy in determining both the location x and the corresponding momentum p_x. In Eqs. (5.4) and (5.13) h_x and h_p were introduced because we did not care to determine the location and momentum with an accuracy better than h_x and h_p, respectively. Now, we must acknowledge that nature imposes upon us a limit on the accuracy with which we can determine *simultaneously* the location and the corresponding momentum. Thus, in Eq. (5.14), h_x and h_p cannot both be arbitrarily small, but their product must be of the order of the Planck constant $h = 6.626 \times 10^{-34}$ J s. Hence, we set

$$h_x h_p \approx h, \tag{5.15}$$

and instead of Eq. (5.14), we write (see Note 2):

$$H_{\max}(\text{location and momentum}) = \log\left(\frac{L\sqrt{2\pi emk_B T}}{h}\right). \tag{5.16}$$

5.2.4 *The SMI of a Particle in a Box of Volume V*

We consider again one simple particle in a cubic box of volume V. We assume that the location of the particle along the three axes x, y, and z is independent. Therefore, we can write the SMI of the location of the particle in a cube of edges L and volume V as

$$H(\text{location in 3D}) = 3H_{\max}(\text{location in 1D}). \tag{5.17}$$

Similarly, for the momentum of the particle we assume that the momentum (or the velocity) along the three axes x, y, and z are independent. Hence, we write

$$H_{\max}(\text{momentum in 3D}) = 3H_{\max}(\text{momentum in 1D}).$$

We combine the SMI's of the locations and momenta of one particle in a box of volume V, taking into account the uncertainty principle. The result is

$$H_{\max}(\text{location and momentum in 3D}) = 3\log\left(\frac{L\sqrt{2\pi e m k_B T}}{h}\right). \tag{5.18}$$

5.2.5 *The SMI's of Locations and Momenta of N Independent Particles in a Box of Volume V*

The next step is to proceed from one particle in a box to N independent particles in a box of volume V. Given the location (x, y, z) and the momentum (p_x, p_y, p_z) of one particle within the box, we say that we know the *microstate* of the particle. If there are N particles in the box, and if their microstates are independent, we can write the SMI of N such particles simply as N times the SMI of one particle, i.e.

$$\text{SMI}(N \text{ independent particles}) = N \times \text{SMI}(\text{one particle}). \tag{5.19}$$

This equation would have been correct if the microstates of all the particles were independent. In reality, there are always correlations between the microstates of all the particles; one is due to *intermolecular interactions* between the particles, and the other is due to the *indistinguishability* between the particles. We will discuss these two sources of correlation separately.

(i) Correlation due to indistinguishability

Recall that the microstate of a single particle includes the location and the momentum of that particle. Let us focus on the location of one particle in a box of volume V. We have written the locational SMI as

$$H_{\max}(\text{location}) = \log V. \tag{5.20}$$

Recall that this result was obtained for the continuous locational SMI. It does not take into account the divergence of the limiting procedure. In order to explain the source of the correlation due to indistinguishability, suppose that we divide the volume V into a very large number of small cells, each of volume V/M. We are not interested in the exact location of each particle, but only in which cell each particle is. The total number of cells is M, and we assume that the total number of particles is $N \ll M$. Suppose that $N = 2$ and $M = 100$, as shown in Fig. 5.1. If each cell can contain at most one particle, then there are M possibilities of putting the first particle in one of the cells, and $M - 1$ possibilities of putting the second particle in the remaining empty cells. Altogether, we have $M(M - 1)$ possible microstates, or configurations, for two particles.

The probability that a particle is found in cell j (or cell i) is

$$\Pr(j) = \Pr(i) = \frac{1}{M}. \tag{5.21}$$

The probability that one particle is found in cell i, and the second in a different cell j, is

$$\Pr(i, j) = \frac{1}{M(M-1)}. \tag{5.22}$$

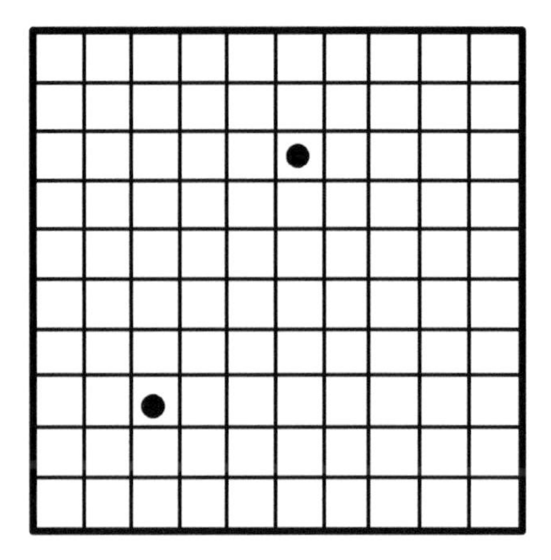

Fig. 5.1. Two particles in 100 cells.

Therefore, we see that even in this simple example there is correlation between the events "one particle in i" and "one particle in j."

$$g(i,j) = \frac{\Pr(i,j)}{\Pr(i)\Pr(j)} = \frac{M^2}{M(M-1)} = \frac{1}{1-\frac{1}{M}}. \tag{5.23}$$

This correlation is easy to understand. For any finite M, the conditional probability of finding a particle in cell i given another particle in cell j is different from the (unconditional) probability of finding the particle in i.

Clearly, this correlation can be made as small as we wish, by taking $M \gg 1$ (or, in general, $M \gg N$). There is another correlation which we cannot eliminate and is due to the indistinguishability of the particles.

Note that in counting the total number of configurations we have implicitly assumed that the two particles are labeled, say red and blue. In this case, we count the two configurations in Fig. 5.2 as *different* configurations: "blue particle in cell i, and red particle in cell j" and "blue particle in cell j, and red particle in cell i."

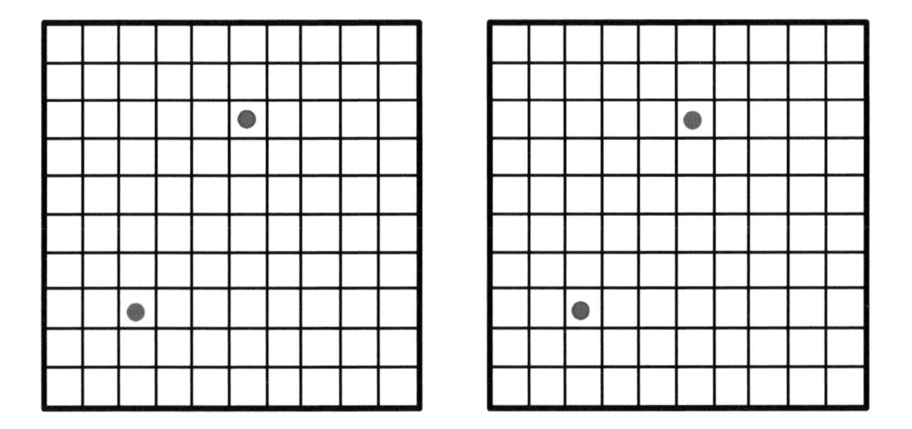

Fig. 5.2. Two distinguishable particles in 100 cells.

Atoms and molecules are indistinguishable by nature; we cannot label them. Therefore, the two microstates (or configurations) in Fig. 5.2 are *indistinguishable*. This means that the total number of configurations is not $M(M-1)$, but

$$\text{Number of configurations} = \frac{M(M-1)}{2} \to \frac{M^2}{2}, \text{ for large } M. \tag{5.24}$$

For very large M, we have a correlation between the events "particle in i" and "particle in j."

$$g(i,j) = \frac{\Pr(i,j)}{\Pr(i)\Pr(j)} = \frac{M^2}{M^2/2} = 2. \tag{5.25}$$

Exercise 5.1

Calculate the correlation function for $N = 3$ and very large M, when the particles are distinguishable and when they are indistinguishable.

For N particles distributed in M cells, we have a correlation function (for $M \gg N$):

$$g(i_1, i_2, \ldots, i_n) = \frac{M^N}{M^N/N!} = N!. \tag{5.26}$$

This means that for N indistinguishable particles we must divide the number of configurations M^N by $N!$. Thus, in general, by removing the "labels" on the particles the number of configurations is *reduced* by $N!$. For two particles, the two configurations shown in Fig. 5.2 reduce to one as shown in Fig. 5.1.

Now that we know that there are correlations between the events "one particle in i_1" "one particle in i_2," ... "one particle in i_n," we can define the MI corresponding to this correlation. We write this as

$$I(1;2;\ldots;N) = \ln N!. \tag{5.27}$$

The SMI for N indistinguishable particles will then be

$$H(N \text{ particles}) = \sum_{i=1}^{N} H(\text{one particle}) - \ln N!. \tag{5.28}$$

(See the definition of the total mutual information in Sec. 4.1.)

Using the SMI for the location and momentum of one particle in Eq. (5.18), we can write the final result for the SMI of N indistinguishable (but noninteracting) particles as

$$H(N \text{ indistinguishable particles}) = N\log V\left(\frac{2\pi mek_BT}{h^2}\right)^{3/2} - \log N!. \tag{5.29}$$

Using the Stirling approximation for $\log N!$ (note again that we use the natural logarithm) in the form

$$\log N! \approx N\log N - N. \tag{5.30}$$

We have the final result for the SMI of N indistinguishable particles in a box of volume V, at temperature T:

$$H(1,2,\ldots,N) = N\log\left[\frac{V}{N}\left(\frac{2\pi mk_BT}{h^2}\right)^{3/2}\right] + \frac{5}{2}N. \tag{5.31}$$

This is a remarkable result. By multiplying the SMI of N particles in a box of volume V at temperature T, by a constant factor (k_B if we use the natural log, or $k_B log_e 2$ if the log is to the base 2), one gets the *entropy* — the *thermodynamic entropy* — of an ideal gas of simple particles. This equation was derived by Sackur (1911) and by Tetrode (1912), by using Boltzmann's definition of entropy.

One can convert this expression into the entropy function $S(E, V, N)$, by using the relationship between the total kinetic energy of the system and the total kinetic energy of all the particles:

$$E = N\frac{m\langle v\rangle^2}{2} = \frac{3}{2}Nk_B T. \tag{5.32}$$

The explicit entropy function of an ideal gas is

$$S(E, V, N) = Nk_B \ln\left[\frac{V}{N}\left(\frac{E}{N}\right)^{3/2}\right] + \frac{3}{2}k_B N\left[\frac{5}{3} + \ln\left(\frac{4\pi m}{3h^2}\right)\right]. \tag{5.33}$$

We can use this equation as a definition of the entropy of a system characterized by constant energy, volume, and number of particles. Note that when we combine all the terms under the logarithm sign, we must get a dimensionless quantity.

(ii) Correlation due to intermolecular interactions

In Eq. (5.32) or (5.33), we got the entropy of a system of noninteracting simple particles (ideal gas). In any real system of particles, there are some interactions between the particles. One of the simplest interaction energy potential functions is shown in Fig. 5.3. Without getting into any details on the function $U(r)$ shown in the figure, it is clear that there are two regions of distances, $0 \leq r \lesssim \sigma$ and $\sigma \leq r \lesssim \infty$, where the slope of $U(r)$ is negative and positive, respectively. A Negative slope corresponds to repulsive forces between the pair of particles when they are at a distance smaller than σ. This is the reason why σ is sometimes referred to as the *effective diameter* of the particles. For larger distances, $r \gtrsim \sigma$, we observe attractive forces between the particles.

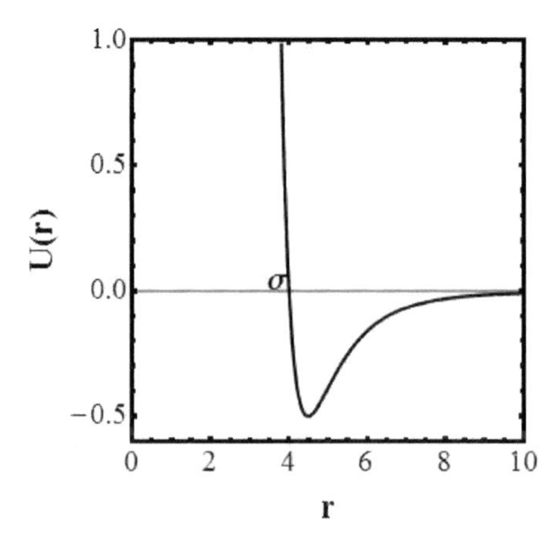

Fig. 5.3. The general form of a pair potential between two particles.

Intuitively, it is clear that interactions between the particles induce correlations between the locational and momentum probabilities of the two particles. For hard-sphere particle, there is an infinitely strong repulsive force between two particles when they approach a distance of $r \leq \sigma$. Thus, if we know the location $\boldsymbol{R}_1$ of one particle, we can be sure that a second particle, at $\boldsymbol{R}_2$, is not in a sphere of diameter σ around the point $\mathbf{R}_1$ [see Fig. 5.4(a)]. This *repulsive* interaction may be said to introduce *negative correlation* between the locations of the two particles.

On the other hand, two argon atoms *attract* each other at distances $r > 4$ Å. Therefore, if we know the location of one particle, say at $\boldsymbol{R}_1$, the probability of observing a second particle at $\boldsymbol{R}_2$ is *larger* than the probability of finding the particle at $\boldsymbol{R}_2$ in the absence of a particle at $\boldsymbol{R}_1$ [see Fig. 5.4(b)]. In this case, we get *positive correlation* between the locations of the two particles.

We can conclude that in both cases (attraction and repulsion) there are correlations between the particles. These correlations can be cast in the form of MI which *reduces* the SMI of a system of N ideal particles in an ideal gas. The mathematical details of these correlations are discussed in Ben-Naim (2008).

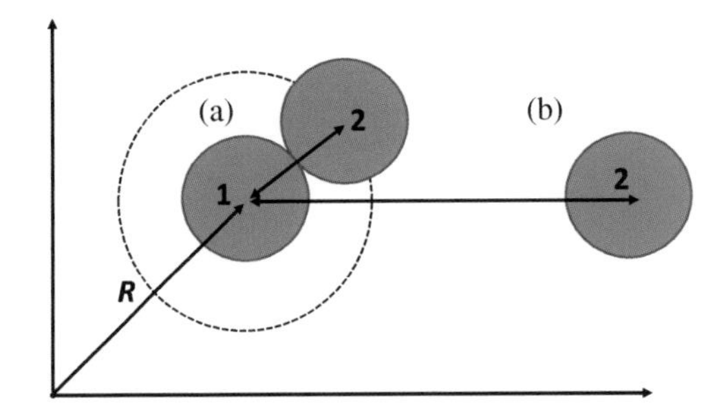

Fig. 5.4. (a) Negative correlation in the region excluded for particle 2, given a particle 1 at R; and (b) positive correlation outside the region excluded for particle 2, given a particle 1 at R.

Here, we show only the form of the MI for the limit of very low densities. At this limit, we can assume that there are only *pair correlations*, and neglect all higher order correlations. The MI due to these correlations is

$$I(\text{due to correlations in pairs}) = \frac{N(N-1)}{2} \times \int P(\mathbf{R}_1, \mathbf{R}_2) \log g(\mathbf{R}_1, \mathbf{R}_2) d\mathbf{R}_1 d\mathbf{R}_2, \tag{5.34}$$

where $P(\mathbf{R}_1, \mathbf{R}_2) d\mathbf{R}_1 d\mathbf{R}_2$ is the probability of finding one particle in $d\mathbf{R}_1$ at $\mathbf{R}_1$ and a second particle in $d\mathbf{R}_2$ at $\mathbf{R}_2$. The correlation function is defined by (see also Sec. 4.5)

$$g(\mathbf{R}_1, \mathbf{R}_2) = \frac{P(\mathbf{R}_1, \mathbf{R}_2)}{P(\mathbf{R}_1) P(\mathbf{R}_2)}. \tag{5.35}$$

Note again that log g can be either positive or negative, but the average in Eq. (5.34) must be positive.

5.2.6 *Conclusion*

We now summarize the main steps leading from the SMI to the entropy.

We started with the SMI associated with the *locations* and *momenta* of the particles. We calculated the distribution of the locations and momenta that *maximizes* the SMI. We referred to this distribution as the *equilibrium distribution*. Let us denote this distribution of the locations and momenta of all the particles by $f_{eq}(\boldsymbol{R}, \boldsymbol{p})$.

Next, we use the equilibrium distribution to calculate the SMI of a system of N particles in volume V and at temperature T. This SMI is, up to a multiplicative constant ($k_B \ln 2$), identical with the *entropy* of an ideal gas at *equilibrium*. This is the reason we referred to the distribution which maximizes the SMI (denoted as f^* in Chap. 2) as the *equilibrium distribution*.

It should be noted that in the derivation of the entropy, we used the SMI twice: first to calculate the distribution that maximizes the SMI, then to evaluate the maximum SMI corresponding to this distribution. The distinction between the concepts of SMI and entropy is essential. Referring to SMI (as many do) as entropy inevitably leads to such an awkward statement: the maximum value of the entropy (meaning the SMI) is the entropy (meaning the thermodynamic entropy). The correct statement is that the SMI associated with locations and momenta is defined for any system — small or large, at equilibrium or far from equilibrium. This SMI, not the entropy, evolves into a maximum value when the system reaches equilibrium. In this state, the SMI becomes proportional to the entropy of the system.

Since the entropy is a special case of an SMI, it follows that whatever interpretation one accepts for the SMI, it will be automatically applied to the concept of entropy. The most important conclusion is that entropy, being a state function, *is not a function of time*. Entropy does not change with time, and entropy does not have a tendency to increase.

We said that the SMI may be defined for a system with any number of particles including the case $N = 1$. This is true for the SMI. When we talk about the entropy of a system, we require that the system be very large. The reason is that only for such systems is the entropy formulation of the second law of thermodynamic valid. This topic is discussed in the next section.

5.3 The Entropy Formulation of the Second Law

In the previous section, we derived and interpreted the concept of entropy. This part answers the question: What is entropy?

Knowing what entropy is leaves the question "Why does entropy always increase?" unanswered. As we will soon see, the statement that "entropy always increases" is not only false, but meaningless, unless one specifies the system and the process for which the entropy changes.

This question is considered to be among the most challenging ones. This property of entropy is also responsible for the mystery surrounding the concept of entropy. In this section, we discuss the origin of the increase in entropy in some specific processes in some specific systems. The correct answer to the question of why entropy "always" increases removes much of the mystery associated with entropy.

In this section, we "derive" the correct *answer* to the correct *questions*: When and why does the entropy of a system increase?

5.3.1 *The Simplest Expansion Process of an Ideal Gas From V to 2V*

Consider the following process. We have a system characterized by E, V, N. (This means N particles, in a volume V having total energy E.) We assume that all the energy of the system is due to the kinetic energy of the particles. We neglect any interactions between the particles, and if the particles have any internal energies (say, vibrational, rotational, electronic, nuclear, etc.), these will not change in the process. We now remove a partition between the two compartments, as in Fig. 5.5, and observe what happens. Experience tells us that once we remove the partition, the gas will expand to occupy the entire system of volume $2V$. Furthermore, if both the initial and the final state

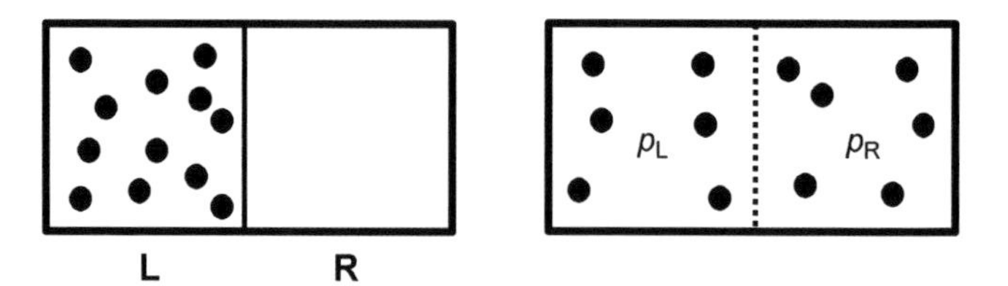

Fig. 5.5. Expansion from V to 2 V.

are equilibrium states, then we can apply the entropy function [either Eq. (5.31) or Eq. (5.33)] to calculate the change in the entropy in this process, i.e.

$$\Delta S(V \to 2V) = Nk_B \ln \frac{2V}{V} = Nk_B \ln 2. \qquad (5.36)$$

Note carefully that this entropy change corresponds to the *difference* in the entropy of the system at two equilibrium states; the initial and the final state (Fig. 5.5).

The informational interpretation of this quantity can be obtained by dividing ΔS by the constant factor $k_B \ln 2$, and we get

$$\Delta H(V \to 2V) = \frac{\Delta S}{k_B \ln 2} = N. \qquad (5.37)$$

This means that the SMI of the system increased by N bits. The reason is simple. Initially, we know that all N particles are in a volume V, and after removal of the partition we lose one bit per particle. We need to ask one question to find out where a particle is: in the right (R) or in the left (L) compartment.

Now that we understand the meaning of this entropy change, we turn to studying the *cause* of this entropy change. Specifically, we ask: Why does the entropy of this process increase?

Before we answer this question we will try to answer a more fundamental question: Why did this process occur at all?

We will see that the answer to the second question leads to the answer to the first question.

Clearly, if the partition separating the two compartments is not removed, nothing will happen; the gas will not expand and the entropy of the system will not change. We can conclude that, having a system characterized by (E, V, N), the entropy is fixed and *will not change with time*.

Let us examine what will happen when we remove the partition separating the two compartments in Fig. 5.5.

Instead of removing the entire partition, we open a small window between the two compartments. This will allow us to follow the process in small steps. If the window is small enough, we can expect that only one particle at the time will pass through it.

Starting with all the N particles in the L compartment, we open the window and observe what will happen.

Clearly, the first particle which crosses the window will be from the L to R compartment, simply because there are no particles in the R compartment.

After some time, there is some chance that a second particle will cross the window. This second particle, with a large probability, will be from L. Clearly, as long as there are many more particles in L than in R, most of the transitions will be from L to R. For a more detailed description, as well as simulation of this process, see Ben-Naim (2008, 2010). The number of particles in R will be denoted by n, and the number in L by $N - n$. The pair of numbers $(N - n, n)$ may be referred to as a distribution of particles in the two compartments. Dividing by N, we get a pair of numbers $(p_L, p_R) = (1 - p, p)$, where $p = n/N$. Clearly, this pair of numbers is a probability distribution $(p_L, p_R \geq 0, p_L + p_L = 1)$. We can refer to it as the temporary probability distribution, or simply the *state distribution*. More precisely, this is the locational state of the particles. Since we have an ideal gas, the energy, the temperature, and the velocity or momentum distribution of the particles will not change in this process.

For each *state distribution* $(1-p,p)$, we can define the corresponding SMI by

$$H(p) = -p\log p - (1-p)\log(1-p). \tag{5.38}$$

Note that p changes with time, and as a result $H(p)$ will also change with time.

If we follow the change of the SMI, we will observe a nearly monotonic increasing function of time as we have seen in Fig. 3.27 [for actual simulations, see Ben-Naim (2008, 2010)]. The larger N is, the more perfect monotonic the curve will be, and once n reaches the value $N/2$ the value of the SMI will stay there "forever." For any N, there will be fluctuations, both on the way up to the maximum and after reaching the maximum. However, for very large N these fluctuations will be unnoticeable. After some time, we reach an equilibrium state. The equilibrium state is reached when the locational distribution is such that it maximizes the SMI, namely

$$p_{\text{eq}} = \frac{N}{2}, \tag{5.39}$$

and the corresponding SMI is

$$H_{\max} = N\left[-\frac{1}{2}\log\frac{1}{2} - \frac{1}{2}\log\frac{1}{2}\right] = N. \tag{5.40}$$

Note again that here we are concerned with the locational distribution with respect to being either in L or in R. The momentum distribution does not change in this process.

Once we reach the equilibrium state, we can ask: What is the *probability* of finding the system such that there is $N-n$ in L and n in the R? Since the probability of finding a specific particle in either L or R is 1/2, the probability of finding the probability distribution

$(N - n, n)$, or the state distribution $(1 - p, p)$, is

$$P(N - n, n) = \frac{N!}{n!(N - n)!} \left(\frac{1}{2}\right)^{N-n} \left(\frac{1}{2}\right)^{n}$$
$$= \binom{N}{n} \left(\frac{1}{2}\right)^{n}. \tag{5.41}$$

It is easy to show that this probability function has a maximum at $n = N/2$. Clearly, if we sum over all n, and use the binomial theorem, we get, as expected,

$$\sum_{n=0}^{N} P(N - n, n) = \sum_{n=0}^{N} \binom{N}{n} \left(\frac{1}{2}\right)^{N} = \left(\frac{1}{2}\right)^{N} 2^{N} = 1. \tag{5.42}$$

We now use the Stirling approximation in the form

$$\ln N! \approx N \ln N - N \tag{5.43}$$

to rewrite Eq. (5.41) as

$$\ln P(1 - p, p) \approx -N\ln 2 - N[(1 - p)\ln(1 - p) + p\ln p] \tag{5.44}$$

or, equivalently, after dividing by ln 2, we get

$$P(1 - p, p) \approx \left(\frac{1}{2}\right)^{N} 2^{N \times H(p)}. \tag{5.45}$$

If we use, instead of the approximation (5.43), the approximation

$$\ln N! \approx N\ln N - N + \frac{1}{2}\ln(2\pi N), \tag{5.46}$$

we get, instead of (5.45), the approximation

$$P(1 - p, p) \approx \left(\frac{1}{2}\right)^{N} \frac{2^{NH(p)}}{\sqrt{2\pi N p(1 - p)}}. \tag{5.47}$$

Note that in general the probability Pr of finding the distribution $(1 - p, p)$ is related to the SMI of that distribution. We now compare the probability of finding the state distribution $\left(\frac{1}{2}, \frac{1}{2}\right)$ with the probability of finding the state distribution (1, 0). From the approximation (5.47), we have

$$P\left(\frac{1}{2}, \frac{1}{2}\right) = \sqrt{\frac{2}{\pi N}}. \tag{5.48}$$

For the state (1,0), we can use the exact expression (5.41):

$$P(1, 0) = \left(\frac{1}{2}\right)^N. \tag{5.49}$$

The ratio of these two probabilities is

$$\frac{P\left(\frac{1}{2}, \frac{1}{2}\right)}{P(1, 0)} = \sqrt{\frac{2}{\pi N}} 2^N. \tag{5.50}$$

Note carefully that $P\left(\frac{1}{2}, \frac{1}{2}\right)$ *decreases* with N. We have already seen this in Figs. 0.6 and 0.7. However, the ratio of the two probabilities in Eq. (5.50) *increases* with N.

The corresponding difference in the SMI is

$$H\left(\frac{1}{2}, \frac{1}{2}\right) - H(1, 0) = N - 0 = N. \tag{5.51}$$

What about the entropy change? Here, one must be very careful. Equations (5.48) and (5.49) were computed for the system *after* we removed the partition (or opened the window). We see that the ratio of the probabilities in Eq. (5.50) is overwhelmingly in favor of the state $\left(\frac{1}{2}, \frac{1}{2}\right)$. To calculate the entropy difference in this process, let us denote by $S_i = S(E, V, N)$ the entropy of the initial state. The entropy of the final state $S_f = S(E, 2V, N)$ may be obtained by multiplying Eq. (5.51) by $k_B \ln 2$, and adding it to the entropy of the initial state S_i:

$$S_f = S_i + (k_B \ln 2)N. \tag{5.52}$$

The change in entropy is therefore

$$\Delta S = S_f - S_i = Nk_B \ln 2, \tag{5.53}$$

which agrees with Eq. (5.36).

It should be emphasized that the ratio of probabilities (5.50) and the difference between the entropies in Eq. (5.53) are computed for *different* states of the same system. In Eq. (5.53), S_f and S_i are the entropies of the system in the *final* and *initial equilibrium* states, respectively. These two equilibrium states are characterized by $(E, 2V, N)$ and (E, V, N), respectively. The corresponding entropies are $S_f = S(E, 2V, N)$ and $S_i = S(E, V, N)$; the latter is the entropy of the system *before* removing the partition. On the other hand, the ratio of the probabilities in Eq. (5.50) is calculated at equilibrium *after* removing the partition.

We can now answer the question posed at the beginning of this section. After the removal of the partition, the gas will expand and attain a new equilibrium state. The *reason* for the change from the initial to the final state is *probabilistic.* The probability of the final state $\left(\frac{1}{2}, \frac{1}{2}\right)$ is overwhelmingly larger than the probability of the initial state $(1, 0)$ *immediately after the removal of the partition.* As a result of the monotonic relationship between the probability $P(1-p, p)$ and the SMI [see Eq. (5.45) or (5.47)], whenever the probability increases, the SMI increases too. At the state for which the SMI is maximum, we can calculate the change in entropy, which is larger by Nk_B ln 2 relative to the entropy of the initial equilibrium state S_i, i.e. *before* the removal of the partition. We can say that the process of expansion occurs *because of* the overwhelmingly larger probability of the final equilibrium state. The increase in the entropy of the system is a *result* of the expansion process, not the *cause* of the process.

A caveat

Quite often, one might find in textbooks Boltzmann's definition of entropy in terms of the number of states:

$$S = k_B \ln W. \tag{5.54}$$

W in this equation is often referred to as a probability. Of course, W cannot be a probability, which by definition is a number between 0 and 1. More careful writers will tell you that the ratio of the number of states is the ratio of the probabilities, i.e. for the final and the initial state, one writes

$$\frac{W_f}{W_i} = \frac{P_f}{P_i}. \tag{5.55}$$

This is true but one must be careful to note that while W_i is the number of states of the system *before* the removal of the partition, both of the corresponding probabilities P_i and P_f pertain to the same system *after* the removal of the partition, i.e. when equilibrium is reached.

Very often, you might find an erroneous statement of the second law based on Eq. (5.54), as follows: the number of states of the system tends to increase, and therefore the entropy tends to increase too. This statement is not true; both W and S in Eq. (5.54) are defined for an equilibrium state, and both do not have a tendency to increase with time!

5.3.2 *The Expansion of an Ideal Gas From V to 3V*

Before we discuss the general formulation of the second law, the following exercise is suggested:

Exercise 5.2

You are shown a two-compartment system as in Fig. 5.6. Unlike the previous case, we remove the partition and we observe an expansion from the initial volume V to $3V$.

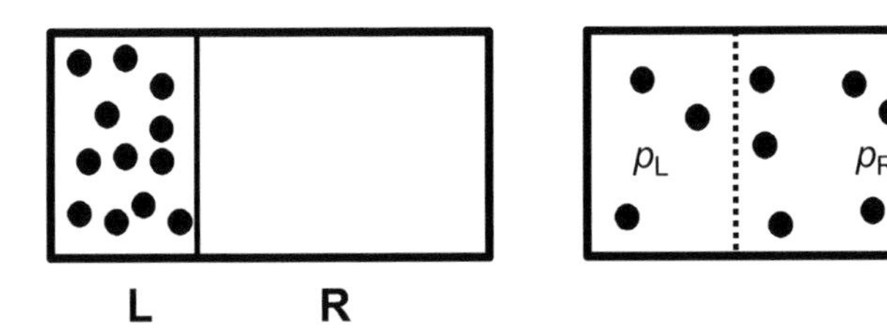

Fig. 5.6. Expansion from V to 3 V.

Calculate the change in entropy in this process, assuming that the system is isolated and there are no interactions between the particles.

Repeat the arguments as in Sec. 5.3.1 regarding the probabilities of the initial and final states, the corresponding change in the SMI, and the entropy. Formulate the second law for this particular process.

5.3.3 *The General Case of a c Compartment System*

We start with N, simple noninteracting particles distributed in c compartments, such that there are $n_i^{(\mathrm{in})}$ particles in the ith compartment; see Fig. 5.7. The total volume of the system is V and the temperature T throughout the system is fixed. For this system, we have the two equalities

$$V = \sum_{i=1}^{c} = V_i \tag{5.56}$$

$$N = \sum_{i=1}^{c} = n_i^{(\mathrm{in})}. \tag{5.57}$$

We also assume that each compartment can accommodate any number of particles, $0 \le n_i \le N$. The vector $\boldsymbol{n}^{(\mathrm{in})} = (n_1^{(\mathrm{in})}, n_2^{(\mathrm{in})}, \ldots, n_c^{(\mathrm{in})})$ may be referred to as a *partition* of the number N into c numbers $n_i^{(\mathrm{in})}$ such that the sum of $n_i^{(\mathrm{in})}$ is equal to N. In this section, we use the word "partition" for the barriers separating the compartments and the vector $\boldsymbol{n}^{(\mathrm{in})}$ will be referred to as the initial state distribution of the system.

Initially, we have a *constrained equilibrium state*. The entropy of the system as a whole is given by

$$S_{\mathrm{total}}^{(\mathrm{in})} = \sum_{i=1}^{c} = S_i^{(\mathrm{in})}, \tag{5.58}$$

where $S_i^{(\mathrm{in})}$ is the entropy of the gas in the compartment i.

We now show that whenever we remove all the partitions between the compartments, the total entropy of the system will increase. This will be the basis on which we will establish the entropy formulation of the second law.

The process is shown in Fig. 5.7. Initially, we have a constrained equilibrium state (before the removal of the partitions); see Fig. 5.7(a). Immediately after the removal of the partitions, the system is as shown in Fig. 5.7(b). At this point, the system is not in an equilibrium state. It is still characterized by the vector $(n_1^{(\mathrm{in})}, n_2^{(\mathrm{in})}, \ldots, n_c^{(\mathrm{in})})$. After some time, the system will reach a new (unconstrained) equilibrium state. We denote by $(n_1^{(\mathrm{f})}, n_2^{(\mathrm{f})}, \ldots, n_c^{(\mathrm{f})})$ the final equilibrium state. It should be noted that $n_i^{(\mathrm{f})}$ is the number of particles in the compartment i, in the *absence* of the partitions (shown as dashed lines in Fig. 5.7).

We already know that the locational SMI for each particle is maximum for the uniform distribution, i.e. the probability of finding a

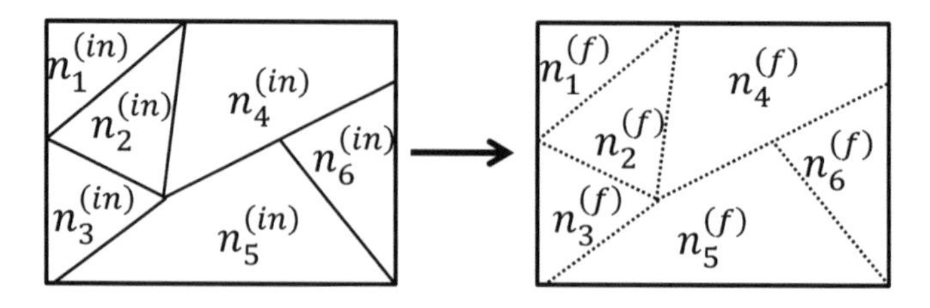

Fig. 5.7. Generalization of the expansion process.

single particle in any element of volume dV at a point $\boldsymbol{R}$ is

$$f(\boldsymbol{R})dV = \frac{dV}{V}. \tag{5.59}$$

$f(\boldsymbol{R})$ is independent of the locational vector $\boldsymbol{R}$.

If we choose dV to be small enough so that at most one center of a particle can be accommodated in this volume, the probability of finding any particle in dV is simply NdV/V. We denote by $\rho = N/V$ the density of particles in the final equilibrium state. The average number of particles in each compartment (in the absence of the partitions) is thus

$$n_i^{(\mathrm{f})} = \rho V_i = \frac{NV_i}{V}. \tag{5.60}$$

Therefore, when we remove the partitions separating all the compartments, the system will move to a new equilibrium state. In this equilibrium state, the uniform locational distribution has the largest probability. Uniform distribution implies that the average number of particles in any compartment is proportional to the volume of that compartment, i.e. $n_i^{(\mathrm{f})} = \rho V_i$ or $x_i^{(\mathrm{f})} = \frac{V_i}{V}$.

The corresponding change in entropy can be calculated from the entropy function (5.33). Assuming that each compartment has initially a large number of particles, we can define the entropy of each compartment, and the difference in entropy for the process in Fig. 5.7 is

$$\Delta S = S(E, V, N) - \sum_{i=1}^{c} (E_i, V_i, n_i^{(\mathrm{in})})$$

$$= Nk_B \ln\left(\frac{V}{N}\alpha\right) - \sum N_i k_B \ln\left(\alpha \frac{V_i}{n_i^{(\mathrm{in})}}\right). \tag{5.61}$$

In this equation, we have included in α all the factors which do not change in the process, particularly $E_i/n_i^{(\mathrm{in})}$, which is proportional to the temperature T. Denoting the *mole fractions* as

$$x_i^{(\mathrm{in})} = \frac{n_i^{(\mathrm{in})}}{N}, \quad x_i^{(\mathrm{f})} = \frac{n_i^{(\mathrm{f})}}{N}, \tag{5.62}$$

we can rewrite Eq. (5.61) as

$$\begin{aligned} \Delta S &= N k_B \sum_{i=1}^{c} x_i^{(\mathrm{in})} \ln\left[\frac{V n_i^{(\mathrm{in})}}{N V_i}\right] \\ &= N k_B \sum_{i=1}^{c} x_i^{(\mathrm{in})} \ln[x_i^{(\mathrm{in})}/x_i^{(\mathrm{f})}] \geq 0. \end{aligned} \tag{5.63}$$

The equality holds if and only if $x_i^{(\mathrm{in})} = x_i^{(\mathrm{f})}$ for all i. The inequality in (5.63) follows from the fact that apart from Boltzmann constant's and the change in the base of the logarithm, this is the Kullback–Leibler (see Sec. 3.3) distance between the two distributions $\mathbf{x}^{(\mathrm{in})}$ and $\mathbf{x}^{(\mathrm{f})}$.

We can conclude this section by answering the following two questions:

(1) Why did a spontaneous process occur?
(2) Why does the entropy increase?

The answer to the first question is straightforward. Accepting the relative frequency interpretation of probability, a system will always spend more time in states having higher probability. When the system is macroscopic (i.e. N of the order of the Avogadro number), the probability of the final equilibrium state is overwhelmingly larger than the probability of any other state, and therefore the system will move with near-certainty to a new equilibrium state.

Regarding the second question, we should first note that the entropy formulation of the second law is valid for an isolated system. Because of the relationship between the probability and the SMI of the distribution, whenever the probability ratio is larger than 1,

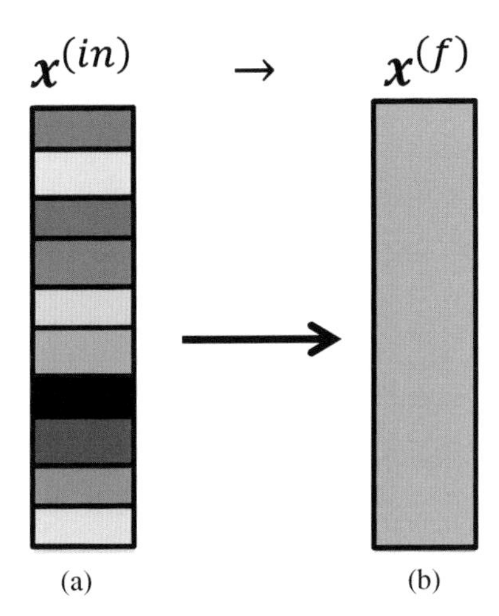

Fig. 5.8. The change of the locational SMI after removal of all the partitions, starting from an initial distribution $\mathbf{x}^{(\mathrm{in})}$ to the uniform distribution $\mathbf{x}^{(\mathrm{f})}$.

the difference in the SMI between the initial and the final state will be positive. Therefore, the difference in the entropy is also positive.

We emphasize again that while the first answer is valid for any spontaneous process in any thermodynamic system, the formulation of the second law in terms of the entropy is valid only for isolated systems.

Let us do a similar experiment as follows (Fig. 5.8): suppose that we start with a system having fixed values of E, V, N. We divide it into c compartments of equal volumes, V/c, each having a different number of particles N_i, such that $\sum N_i = N$. The initial entropy of each compartment is denoted by $S_i^{(\mathrm{in})}$, and the total entropy of the system, assuming that each compartment is macroscopic and at equilibrium, is

$$S^{(\mathrm{in})}(E, V, N; \mathbf{x}^{(\mathrm{in})}) = \sum_{i=1}^{c} S_i^{(\mathrm{in})}, \tag{5.64}$$

where $\mathbf{x}^{(\mathrm{in})}$ is the distribution defined by its components: $x_i^{(\mathrm{in})} = \frac{N_i}{N}$. [See Fig. 5.8(a).]

When we remove all the partitions between the compartments, keeping E, V, N constant, the system will evolve into a new equilibrium state having an entropy value which is *larger* than the initial entropy, and for which the distribution of particles is uniform, i.e. $x_i^{(\mathrm{f})} = 1/c$ [Fig. 5.8(b), and

$$S^{(\mathrm{f})}(E, V, N; \mathbf{x}^{(\mathrm{f})}) \geq S^{(\mathrm{in})}(E, V, N; \mathbf{x}^{(\mathrm{in})}). \tag{5.65}$$

Again, we emphasize that the reason for the evolution of the system, from the initial to the final state,

$$\mathbf{x}^{(\mathrm{in})} \rightarrow \mathbf{x}^{(\mathrm{f})}, \tag{5.66}$$

is probabilistic, and the change in entropy is always positive (under the condition that the system is isolated).

Note that in this example, starting from *any* initial distribution $\mathbf{x}^{(\mathrm{in})}$, the system will always evolve toward the *uniform distribution* $\mathbf{x}^{(\mathrm{f})}$. The reason is that the uniform distribution is the one that maximizes the probability $\Pr(\mathbf{x})$ or, equivalently, the corresponding SMI, $H(\mathbf{x})$. In the next experiment, we show that the uniform distribution does not always characterize the equilibrium distribution. Note that if there are interactions the final locational distribution does not have to be uniform. For instance, if there is more than one phase at equilibrium, the locational distribution will not be uniform. We can now make a general statement regarding the entropy formulation of the second law:

Starting with any constrained equilibrium state of an isolated system, the total entropy of the system will either increase or remain unchanged when we remove any of the constraints. This is the meaning of "Entropy always increases," with which we started this section.

5.3.4 A Process of Expansion in a Gravitational Field at Constant Temperature

We now modify the experiment shown in Fig. 5.8. We have again a system characterized by (T, V, N), and initially it is divided into c

compartments of equal volume, numbered from the bottom $i = 1$ to $i = c$, and with any arbitrary initial distribution $\mathbf{x}^{(\text{in})}$. The total system is in a gravitational field, and therefore the energy of each compartment is

$$E_i = N_i\left(\frac{1}{2}m\langle v\rangle i^2\right) + N_i mgh_i.$$

The total energy of the system is the sum of the kinetic energy EK and the potential energy EP, Thus,

$$E = \sum E_i = EK + EP, \tag{5.67}$$

where $\frac{1}{2}m\langle v\rangle i^2$ is the average kinetic energy of the particles at elevation h_i, m the mass of a particle (one component is assumed), g the gravitational acceleration, and h_i the height of the ith compartment relative to the compartment $i = 1$ at the bottom of the system; see Fig. 5.9. We also assume that in each compartment the potential energy of a particle is constant, mgh_i, but still each compartment is macroscopic and contains a large number of particles. Also, we assume, as before, that there are no interactions between the particles (ideal gas).

If we remove the partitions between all the compartments, the system will evolve from the initial distribution $\mathbf{x}^{(\text{in})}$ to the final

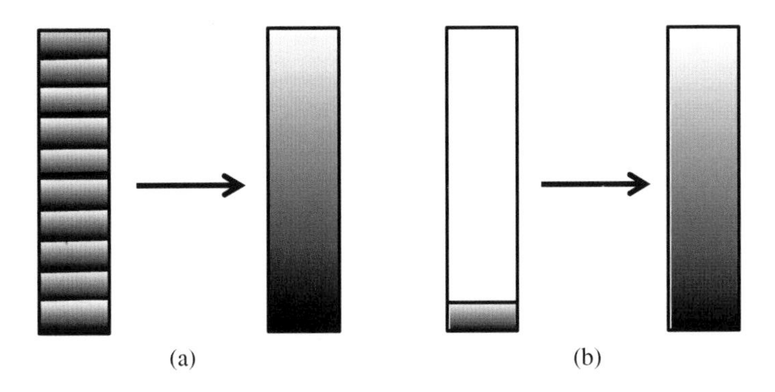

Fig. 5.9. Two processes in a gravitational field at constant T, V, N.

distribution, which is now *nonuniform*, i.e.

$$\mathbf{x}^{(\text{in})} \to \mathbf{x}^{(\text{f})}, x_i^{(\text{f})} = \frac{\exp(-\beta m g h_i)}{\sum_{i=1}^{c} \exp(-\beta m g h_i)}. \tag{5.68}$$

Here, the final distribution is the exponential or Boltzmann distribution, where $\beta = 1/T$, with T being the temperature of the system. Note that since T is constant in the system, the average kinetic energy will also be the same at any level i. [For details, see Ben-Naim (1992).]

Here, we can make the following statements regarding the evolution of the system:

(1) Starting from *any* initial distribution $\mathbf{x}^{(\text{in})}$, the final distribution will be $\mathbf{x}^{(\text{f})}$, as in Eq. (5.68).

(2) The locational SMI of the system might increase, decrease, or remain unchanged:

   (a) If $\mathbf{x}^{(\text{in})}$ is such that $H(\mathbf{x}^{(\text{in})}) > H(\mathbf{x}^{(\text{f})})$, then the total locational SMI will decrease. For instance, if we start with, say, 0.1 mole of particles in each cell, then the locational SMI must decrease toward the Boltzmann distribution $\mathbf{x}^{(\text{f})}$; see Fig. 5.9(a).

   (b) If $\mathbf{x}^{(\text{in})}$ is such that $H(\mathbf{x}^{(\text{in})}) < H(\mathbf{x}^{(\text{f})})$, then the total locational SMI will increase. For instance, starting with 1 mole of particles at level $i = 1$ (or any other level, $i \neq 1$), the locational SMI will increase; see Fig. 5.9(b).

   (c) If $\mathbf{x}^{(\text{in})}$ is the Boltzmann distribution equal to $\mathbf{x}^{(\text{f})}$ in Eq. (5.68), then there will be no change in the locational SMI.

All these conclusions follow from the fact that the distribution which maximizes the SMI also maximizes the probability Pr.

Thus, we see that the SMI of the system does not necessarily increase in any spontaneous process. This is one more argument for making a clear-cut distinction between SMI and entropy. This SMI can either increase or decrease, depending on the initial distribution. The entropy of the system is related to the maximum SMI of the locations and the momentum of the particles. In the processes in

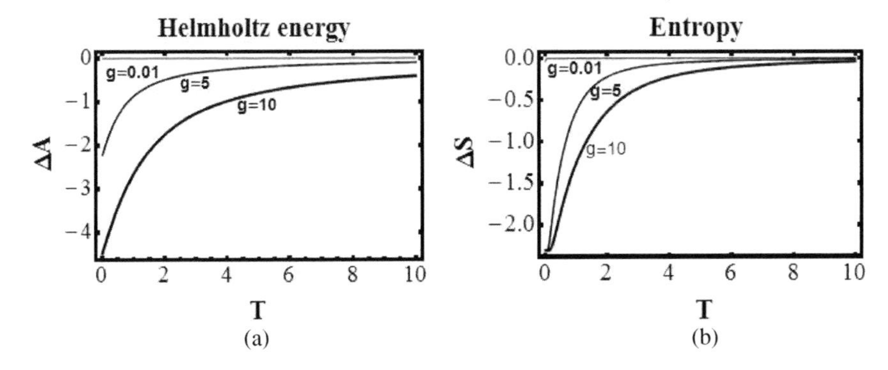

Fig. 5.10. The change in the Helmholtz energy and the entropy for the process in Fig. 5.9(a) as a function of T for different values of g.

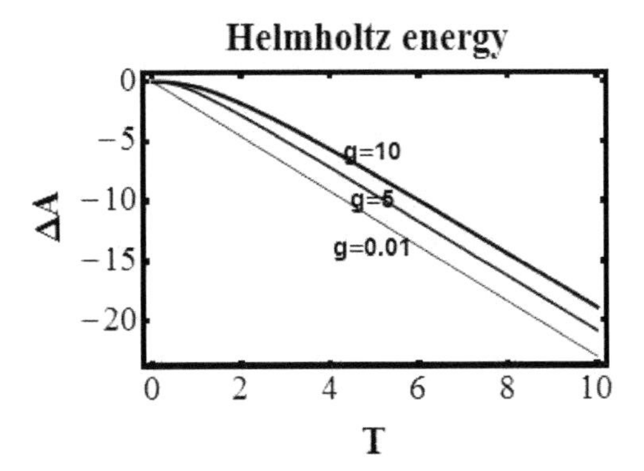

Fig. 5.11. The change in the Helmholtz energy for the process in Fig. 5.9(b) as a function of T for different values of g.

Fig. 5.9, the Helmholtz energy, defined by $A = E - TS$, must decrease. Since the total kinetic energy of the system does not change, only the locational distribution changes in these experiments. The entropy can increase or decrease or remain unchanged.

Figure 5.10 shows the change in Helmholtz energy ΔA, the entropy change ΔS for the process in Fig. 5.9(a). As we can see, both the Helmholtz energy and the entropy *decrease* in the process.

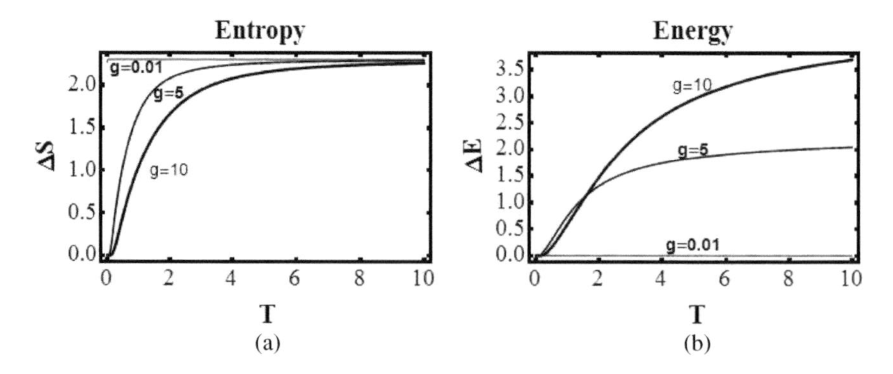

Fig. 5.12. The change in the entropy and in the energy for the process in Fig. 5.9(b) as a function of T for different values of g.

The energy is almost unchanged in this process, $\Delta E \approx 0$. On the other hand, Figs. 5.11 and 5.12 show that while the Helmholtz energy decreases, the entropy and the energy increase in the process of expansion in Fig. 5.9(b).

The reason is that the locational SMI is larger in this process (the final distribution is more widely spread than the initial one) and the average potential energy of the particles increases.

Exercise 5.3

(Expansion in a gravitational field.)

Consider the following two experiments on the expansion of an ideal gas from an initial volume V to a final volume $2V$.

In one experiment, the gas expands from a lower to an upper compartment. In the second, the expansion is downward; see Fig. 5.13(a) and 5.13(b). Estimate qualitatively the change in entropy in these two experiments; for the first, when the entire process is carried out under constant temperature and volume; for the second, when the whole system is isolated.

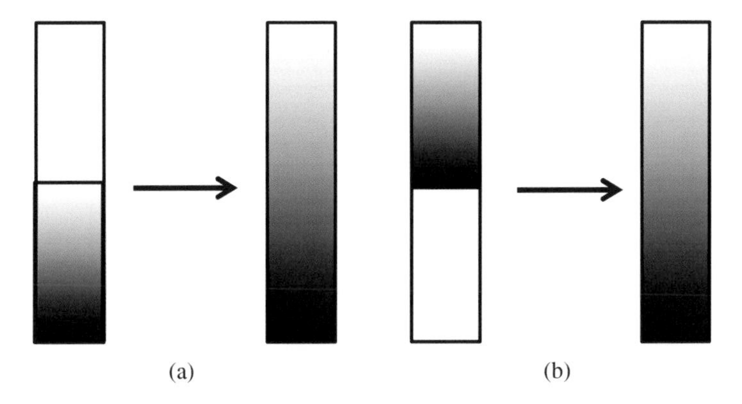

Fig. 5.13. Two processes of expansion in a gravitational field at constant T, V, N: (a) Upward and (b) downward.

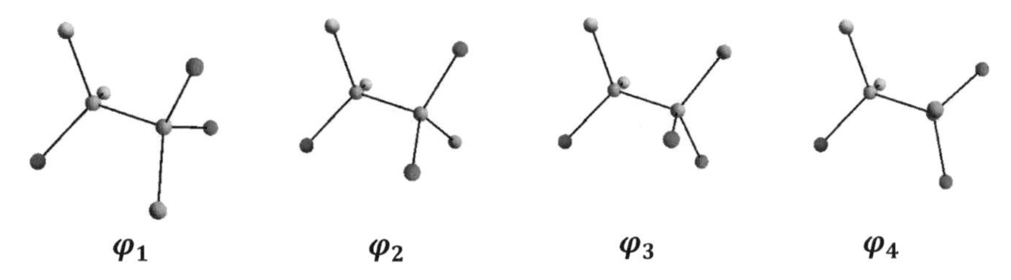

Fig. 5.14. A molecule having different conformations characterized by an angle φ_i.

5.3.5 *Particles Having Internal Rotational Degrees of Freedom*

In this and the next section, we describe two modifications of the experiment discussed in Sec. 5.3.3. In each case, we start with N particles, at some initial temperature T and in a box of volume V.

In the first experiment, we have molecules with different conformations, defined by an angle φ, the angle of rotation around a single bond; see Fig. 5.14 and Fig. 5.15. We assume that there are only c possible angles, and also that there is a catalyst in the presence of

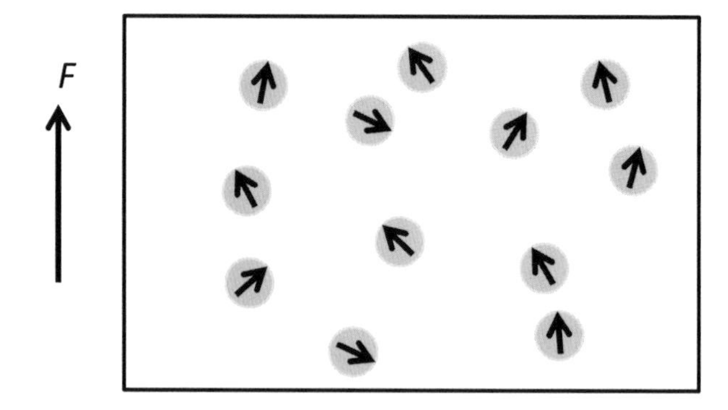

Fig. 5.15. Molecules whose dipole moment which has only a finite number of orientations, in a field F.

which molecules can change their angle of rotation (equivalently, we can assume the existence of an *inhibitor*, in the presence of which no transition between the conformations is possible). As in Sec. 5.3.3, we can prepare the system in an initial state in which there are φ_i molecules in state i, which is characterized by the angle φ_i, and by some internal energy ε_i. The absence of a catalyst, or the presence of an inhibitor, is the analog of the partitions between the compartments in Sec. 5.3.3. The total energy of the system is, as in Eq. (5.64),

$$E_T = \sum_{i=1}^{c} N_i E_i = \sum N_i \left(\frac{1}{2} m \langle v \rangle^2 + \varepsilon_i\right) = EK + EP, \qquad (5.69)$$

where EK is the total kinetic energy of the molecules, and EP the total potential (or internal) energy of the molecules. Here, we assume that the average velocity of all the molecules is independent of their internal state.

Starting with any initial distribution of molecules $\mathbf{x}^{(\mathrm{in})} = (x_1^{(\mathrm{in})}, x_2^{(\mathrm{in})}, \ldots, x_c^{(\mathrm{in})})$, in the absence of a catalyst (or the presence of an inhibitor), the system is viewed as a mixture of c components. The entropy of the system is well defined.

We now add a catalyst (or remove an inhibitor) such that transition between the various states of the molecules is possible. The system will proceed to a new distribution, which we denote by $x^{(f)} = (x_1^{(f)}, x_2^{(f)}, \ldots, x_c^{(f)})$.

In analogy with the case of a system in a gravitational field, here the final distribution will be determined by the internal energies of the molecules, i.e.

$$x_i^{(f)} = C\exp(-\beta\varepsilon_i), \tag{5.70}$$

where $\beta = 1/T$. If the system evolves at constant temperature, then the entropy change in this process can be either positive or negative. The Helmholtz energy, defined by $A = E - TS$, must decrease. Consider the following two extreme cases:

(1) If the initial distribution is uniform, i.e. all $x_i^{(in)} = 1/c$, then adding the catalyst will cause a change in the distribution toward the one defined in Eq. (5.70), which has a lower SMI. Therefore, in this case, the entropy change will be *negative*.
(2) If the initial distribution is such that all molecules are in a single state k, i.e. $x_k^{(in)} = 1$ and $x_i^{(in)} = 0$ for all $i \neq k$, then obviously the initial SMI associated with the conformation of the molecule will be zero. Hence, the entropy change in the process will be *positive*.

5.3.6 *Dipoles in an External Field*

As a final example, consider a system of N molecules at a given temperature T and volume V. Each molecule has a dipole moment (electric or magnetic), which interacts with an external field (electric or magnetic), in such a way that orientation in the direction of the field is more favored.

As in the previous example, we can start with any initial distribution of angles where the angle φ_i is with respect to the external field. As in the previous example also, the system will always evolve to the final equilibrium distribution, as in Eq. (5.70), but now the potential energy of each molecule is determined by the strength of the interaction of the dipole with the external field.

Again, if we let the system evolve under constant temperature, the entropy change will be either positive or negative, depending on the initial distribution. The Helmholtz energy, defined by $A = E - TS$, must decrease. [For a proof, see Ben-Naim (2013, 2016b).]

5.4 The Helmholtz Energy Formulation of the Second Law

In the previous sections, we saw a few examples where, upon the removal of a constraint, the entropy can go either up or down. In such examples, the entropy formulation of the second law does not hold. Instead, the Helmholtz energy formulation of the second law takes over. This formulation states that whenever we start with a constrained equilibrium in a system characterized by the variables T, V, N and remove the constraints, the Helmholtz energy function, $A = E - TS$, will *decrease*. ΔS in such processes can be either positive or negative, but ΔA must be negative.

As we have discussed in connection with the entropy formulation, underlying the Helmholtz energy formulation is a more fundamental principle. Upon the removal of the constraints, the system will evolve from a state of lower probability to a state of higher probability. The ratio of these two probabilities is related to the difference between the Helmholtz energies, i.e.

$$\frac{P(\text{final})}{P(\text{initial})} = \exp(-\beta\Delta A), \tag{5.71}$$

where $\Delta A = A(\text{final}) - A(\text{initial}) < 0$. [For details, see Ben-Naim (1992).]

5.5 The Gibbs Energy Formulation of the Second Law

Here, we briefly formulate the second law for a spontaneous process under constant temperature, pressure, and total number of particles. [For details, see Ben-Naim (1992).]

Starting with any initial constrained equilibrium state, all the thermodynamic functions are well defined, including the internal energy, the entropy, and the Helmholtz energy.

Removing the constraint, keeping T, P, N constant, will always result in a decrease in the Gibbs energy function, defined by $G = A + PV$, where A is the Helmholtz energy, P the pressure, and V the volume of the system.

Again, we note that underlying this formulation of the second law is a more fundamental principle. As in Eq. (5.71), the ratio of the probabilities is now related to the difference in the Gibbs energies, i.e.

$$\frac{P(\text{final})}{P(\text{initial})} = \exp(-\beta \Delta G), \tag{5.72}$$

where $\Delta G = G(\text{final}) - G(\text{initial}) < 0$.

5.6 Summary of Chapter 5

In this chapter, we have seen that the main "driving force" for any spontaneous process in a thermodynamic system is probability. The ratio of the probabilities in the initial and final states is related to the entropy change in an isolated system, to the Helmholtz energy change in a T, V, N system, and to the Gibbs energy change in a T, P, N system.

We have discussed only a one-component system, but the various formulations of the second law also apply to the multicomponent systems in which chemical reactions occur.

In all the formulations, we start with some initial distribution $\mathbf{x}^{(\text{in})}$, and the system will evolve with a probability of nearly 1 to the final distribution $\mathbf{x}^{(\text{f})}$.

Final Exercise (which only a demon can solve)

You are shown a system of N simple noninteracting particles, contained in a box of volume V and at temperature T. You are told that the exact entropy of this system is S_0.

I am a demon and I can reduce my size to the same size as the particles in the system. I get into the box without being noticed by any of the particles (remember that the system is an ideal gas and the particles do not interact with each other). I look around and record the locations and velocities of all particles. I then get out of the box and tell you the exact configuration of the system I have recorded.

Should you revise the value of the entropy of the system after having received the *information* on its precise configuration?

Would the entropy be higher or lower, knowing the exact microstate of the system?

Appendix A

Proof of an Equivalent Markovian Property

Given the Markovian property

$$P(X_n|X_{n-1}, X_{n-2}, \ldots, X_1) = P(X_n|X_{n-1}), \tag{A.1}$$

for any n, we prove that

$$\begin{aligned} &P(X_1, X_2, \ldots, X_{k-1}, X_{k+1}, \ldots, X_n|X_k) \\ &\quad = P(X_1, X_2, \ldots, X_{k-1}|X_k)P(X_{k+1}, \ldots, X_n|X_k). \end{aligned} \tag{A.2}$$

Colloquially, the property (A.1) may be stated as follows: the outcome of the experiment X_n depends only on the "near past," and not on the "distant past." Likewise, we can state (A.2) as follows: *given* the "present" (X_k), the "past" ($X_1, \ldots, X_{k-1}$) and the "future" ($X_{k+1}, \ldots, X_n$) become independent.

To prove the equivalence between the two properties, we first show that

$$\begin{aligned} &P(X_{n+m}, X_{n+m-1}, \ldots, X_{n+1}|X_n, \ldots, X_1) \\ &\quad = P(X_{n+m}, \ldots, X_{n+1}|X_n). \end{aligned} \tag{A.3}$$

For any $m > 1$, this is similar to the Markovian property, but for this case we have several rv's on the left hand side of the vertical line.

To show (A.3), we start with the left hand side of the equation and expand it as follows:

$$\begin{aligned}
&P(X_{n+m},\dots,X_{n+1}|X_n,\dots,X_1)\\
&\quad= P(X_{n+m}|X_{n+m-1},\dots,X_{n+1},X_n,\dots,X_1)\\
&\qquad\times P(X_{n+m-1}|X_{n+m-2},\dots,X_1)\cdots P(X_{n+1}|X_n,\dots,X_1).
\end{aligned} \tag{A.4}$$

Because of the Markovian property, we can delete only part of the "distant past," i.e. all the conditions $(X_{n-1},\dots,X_1)$ on the right hand side of Eq. (A.4), and get

$$\begin{aligned}
&P(X_{n+m},\dots,X_{n+1}|X_n,\dots,X_1)\\
&\quad= P(X_{n+m}|X_{n+m-1},\dots,X_n)\\
&\qquad\times P(X_{n+m-1}|X_{n+m-2},\dots,X_1)\cdots P(X_{n+1}|X_n)\\
&\quad= P(X_{n+m},X_{n+m-1},\dots,X_{n+1}|X_n).
\end{aligned} \tag{A.5}$$

To prove that Eq. (A.2) follows from Eq. (A.1), we start with the right hand side of the former:

$$\begin{aligned}
&P(X_1,X_2,X_{k-1},X_{k+1}\cdots X_n|X_k)\\
&\quad= \frac{P(X_1,X_2,\dots,X_n)}{P(X_k)}\\
&\quad= \frac{P(X_n,\dots,X_{k+1}|X_k,\dots,X_1)P(X_k,\dots,X_1)}{P(X_k)}.
\end{aligned} \tag{A.6}$$

Now, we use the property (A.1) to rewrite Eq. (A.6) as

$$\begin{aligned}
&P(X_1,X_2,X_{k-1},X_{k+1}\cdots X_n|X_k)\\
&\quad= P(X_n,\dots,X_{k+1}|X_k)P(X_{k-1},X_{k-2},\dots,X_1|X_k).
\end{aligned} \tag{A.7}$$

This is the result (A.2), i.e. given X_k the two sequences $(X_1,\dots,X_{k-1})$ and $(X_{k+1},\dots,X_n)$ are independent.

To prove that Eq. (A.1) follows from Eq. (A.2), we start with the left hand side of the former, and use the property (A.2) to obtain

$$\begin{aligned}
&P(X_n, |X_{n-1}, \ldots, X_1) \\
&\quad = \frac{P(X_n, X_{n-1}, \ldots, X_1)}{P(X_{n-1}, \ldots, X_1)} \\
&\quad = \frac{P(X_n, X_{n-2}, \ldots, X_1 | X_{n-1}) P(X_{n-1})}{P(X_{n-1}, \ldots, X_1)} \\
&\quad = \frac{P(X_n | X_{n-1}) P(X_{n-2}, \ldots, X_1 | X_{n-1}) P(X_{n-1})}{P(X_{n-1}, \ldots, X_1)} \\
&\quad = \frac{P(X_n | X_{n-1}) P(X_{n-1}, \ldots, X_1)}{P(X_{n-1}, \ldots, X_1)} = P(X_n | X_{n-1}), \qquad \text{(A.8)}
\end{aligned}$$

which is the required Markovian property (A.1).

Appendix B

Proof of the Uniqueness of the Function H

There are several sets of assumptions leading to the SMI. In this appendix, we follow the assumptions and the proof given by Ash (1965).

For any finite probability distribution, $p_1, \ldots, p_n$, a measure of uncertainty denoted as $H(p_1, \ldots, p_n)$ associated with this distribution is presumed to fulfill the following conditions (referred to as axioms by Ash). The plausibility of these assumptions for a measure of information was discussed in Chap. 2.

(1) If the distribution is uniform, i.e. $p = p_i = 1/n$ for $i = 1, 2, \ldots, n$, the expected function $H(1/n, \ldots, 1/n)$ is a monotonically increasing function of n.
(2) Denote $f(n) = H(\frac{1}{n}, \ldots, \frac{1}{n})$. The function $f(n)$ is expected to be *additive*, i.e.

$$f(n \times m) = f(n) + f(m). \tag{B.1}$$

(3) For any other distribution, the function $H(p_1, \ldots, p_n)$ is expected to have the grouping property. For any partition of the distribution $(p_1, \ldots, p_n) = (p_1, \ldots, p_r, p_{r+1}, \ldots, p_n)$, we define the

probabilities

$$p_A = \sum_{i=1}^{r} p_i, \quad p_B = \sum_{i=n+1}^{r} p_i. \tag{B.2}$$

Then

$$H(p_1, \ldots, p_n) = H(p_A, p_B) + p_A H\left(\frac{p_1}{p_A}, \frac{p_2}{p_A}, \ldots, \frac{p_r}{p_A}\right) + p_B\left(\frac{p_{r+1}}{p_B}, \ldots, \frac{p_n}{p_B}\right), \tag{B.3}$$

for any $r = 1, 2, \ldots, n-1$.
(4) The function $H(p, 1-p)$ is expected to be a continuous function of p.

With these four assumptions, we prove that the only function satisfying these conditions is

$$H(p_1, \ldots, p_n) = -C\sum_{i=1}^{n} p_i \log p_i, \tag{B.4}$$

where C is any positive number, and the logarithm base is an arbitrary number greater than 1.

The proof is carried out in four steps:
(a) The function $f(n)$, defined by $H\left(\frac{1}{n}, \ldots, \frac{1}{n}\right)$ (i.e. n equally probable outcomes), has the property

$$f(n^k) = kf(n), \tag{B.5}$$

for any positive integers n and k.

We prove this by mathematical induction.

For any fixed positive number n, Eq. (B.5) is true for $k = 1$. Assuming that this equality is true for all integers up to $k-1$, it follows from the property (2) that

$$f(n^k) = f(n \times n^{k-1}) = f(n) + f(n^{k-1}). \tag{B.6}$$

According to the induction assumption, we have

$$f(n^{k-1}) = (k-1)f(n). \tag{B.7}$$

Therefore, from Eqs. (B.6) and (B.7), it follows that

$$f(n^k) = kf(n). \tag{B.8}$$

(b) The function $f(n)$ has the form

$$f(n) = C \log n, \tag{B.9}$$

for any integer $n \geq 1$.

For $n = 1$, $\log n = 0$, and from the assumption (2),

$$f(1) = f(1 \times 1) = f(1) + f(1). \tag{B.10}$$

Therefore, $f(1) = 0$. Hence, Eq. (B.9) is true for $n = 1$. For any positive integer $n \geq 2$, we can always find a positive integer k such that for a positive integer r we have the inequality

$$n^k \leq 2^r \leq n^{k+1}. \tag{B.11}$$

Note that for fixed $n \geq 2$ the numbers n^k (with positive k) are larger than n and tend to infinity when $k \to \infty$. Thus, in fact, the inequality (B.11) is valid for any real positive number $r \geq 1$.

By the condition (1), $f(n)$ is a monotonically increasing function of n, and therefore we can write

$$f(n^k) \leq f(2^r) \leq f(n^{k+1}), \tag{B.12}$$

and according to the result (B.5) we can write

$$kf(n) \leq rf(2) \leq (k+1)f(n). \tag{B.13}$$

Dividing by $rf(n)$, we get

$$\frac{k}{r} \leq \frac{f(2)}{f(n)} \leq \frac{k+1}{r}. \tag{B.14}$$

We know that the function $\log n$ is also a monotonically increasing function of n (for any base larger than 1). Hence, we can also write an inequality similar to (B.14) as

$$\frac{k}{r} \leq \frac{\log 2}{\log n} \leq \frac{k+1}{r}. \tag{B.15}$$

Now, we see that both the ratio $f(2)/f(n)$ and the ratio $\log 2/\log n$ are bound between k/r and $(k+1)/r$. Therefore, the distance between these two ratios is bound by

$$\left|\frac{f(2)}{f(n)} - \frac{\log 2}{\log n}\right| \leq \frac{1}{r}, \tag{B.16}$$

for any positive integer r.

The inequality (B.16) is valid for a fixed n. We let $r \to \infty$ and we get the equality

$$\frac{f(2)}{f(n)} - \frac{\log 2}{\log n} = 0 \tag{B.17}$$

or, equivalently,

$$\frac{f(n)}{\log n} = \frac{f(2)}{\log 2} = C. \tag{B.18}$$

Hence,

$$f(n) = C \log n, \tag{B.19}$$

where C is a positive number.

(c) Next, we show that for any rational number, $p = r/s$, the equality

$$H(p, 1-p) = -p\log p - (1-p)\log(1-p) \tag{B.20}$$

holds. To prove this, we use the condition (3) and write

$$f(s) = H\left(\frac{1}{s}, \frac{1}{s}, \dots, \frac{1}{s}\right) = H\left(\frac{r}{s}, \frac{s-r}{s}\right) + \frac{r}{s}f(r) + \frac{s-r}{s}f(s-r), \tag{B.21}$$

where we have simply divided the s equally probable events into two groups of r and $s-r$ events. We now denote $p = r/s$ and apply the result (B.19) to rewrite Eq. (B.21) as

$$C \log s = H(p, 1-p) + Cp \log r + C(1-p)\log(s-r). \qquad \text{(B.22)}$$

Rearranging Eq. (B.22), we get

$$\begin{aligned} H(p, 1-p) &= -C[p \log r - \log s + (1-p)\log(s-r)] \\ &= -C[p \log r - (p+1-p)\log s + (1-p)\log(s-r)] \\ &= -C[p \log p + (1-p)\log(1-p)]. \end{aligned} \qquad \text{(B.23)}$$

We proved Eq. (B.23) for any rational number $p = r/s$. From the assumption of continuity, the validity of Eq. (B.23) for any *real* number $0 \leq p \leq 1$ follows.

(d) So far, we have found the form of H for the uniform distributions, and for binary distribution. We now extend the proof for the case of any distribution of n outcomes. We proceed by mathematical induction. We know the form of H for the cases $n = 1$ and $n = 2$. Now, we assume that

$$H(p_1, \ldots, p_n) = -C \sum p_i \log p_i \qquad \text{(B.24)}$$

is valid for some $n > 2$, and we show that it is valid for $n + 1$. We denote $p_A = \sum_{i=1}^{n} p_i$ and $p_B = p_{n+1}$. Following the grouping property, we have

$$\begin{aligned} H(p_1, \ldots, p_n, p_{n+1}) &= H(p_A, 1-p_A) \\ &\quad + p_A H\left(\frac{p_1}{p_A}, \ldots, \frac{p_n}{p_A}\right) + (1-p_A)H(1) \\ &= -C\left[p_A \log p_A + (1-p_A)\log(1-p_A)\right. \\ &\quad \left. + p_A \sum_{i=1}^{n} \frac{p_i}{p_A} \log \frac{p_i}{p_A}\right] \end{aligned}$$

$$
\begin{aligned}
&= -C\left[\sum_{i=1}^{n} p_i \log\left(\sum_{i=1}^{n} p_i\right) + p_{n+1}\log p_{n+1}\right. \\
&\quad \left. + \sum_{i=1}^{n} p_i \log\frac{p_i}{p_A}\right] \\
&= -C\left[p_{n+1}\log p_{n+1} + \sum_{i=1}^{n} p_i \log p_i\right] \\
&= -C\left[\sum_{i=1}^{n+1} p_i \log p_i\right]. \qquad \text{(B.25)}
\end{aligned}
$$

This completes the proof of the uniqueness of the function H for any finite number of outcomes.

Appendix C

The SMI for the Continuous Random Variable

In this appendix, we discuss the passage to the limit of a continuous distribution. We assume that the rv X can attain any value within the interval (a, b), and that there exists a probability density $f(x)$, such that

$$\Pr(x_1 \leq X \leq x_2) = \int_{x_1}^{x_2} f(x)dx \tag{C.1}$$

and

$$\int_a^b f(x)dx = 1. \tag{C.2}$$

We now divide the interval (a, b) into n intervals, each of size $\delta = (b - a)/n$. We denote the points.

$$x_1 = a, \quad x_1 = a + (i - 1)\delta, \quad x_{n+1} = a + n\delta = b. \tag{C.3}$$

Thus, the quantity

$$P(i, n) = \int_{x_i}^{x_{i+1}} f(x)dx \tag{C.4}$$

is the probability of finding the value of the rv between x_i and x_{i+1}, for a given subdivision into n intervals.

The SMI associated with a fixed n is

$$H(n) = -\sum_{i=1}^{n} P(i,n)\log P(i,n). \tag{C.5}$$

Clearly, since $H(n)$ is defined for a finite value of n, there is no problem in using Eq. (C.5) for any fixed n.

Substituting Eq. (C.4) into Eq. (C.5), we have

$$\begin{aligned} H(n) &= -\sum_{i=1}^{n}\left[\int_{x_i}^{x_{i+1}} f(x)dx\right]\log\left[\int_{x_i}^{x_{i+1}} f(x)dx\right] \\ &= -\sum_{i=1}^{n}\left[\bar{f}(i,n)\delta\right]\log\left[\bar{f}(i,n)\,\delta\right] \\ &= -\sum_{i=1}^{n}\left[\frac{b-a}{n}\bar{f}(i,n)\right]\log\left[\bar{f}(i,n)\right] - \left[\sum_{i=1}^{n}\bar{f}(i,n)\frac{b-a}{n}\right][\log\delta], \end{aligned} \tag{C.6}$$

where $\bar{f}(i,n)$ is some value of the function $f(x)$ between $f(x_i)$ and $f(x_{i+1})$, for a specific value of n. When $n\to\infty$, we have

$$\lim_{n\to\infty}\sum_{i=1}^{n}\bar{f}(i,n)\frac{b-a}{n} = \int_a^b f(x)dx = 1, \tag{C.7}$$

$$\lim_{n\to\infty}\sum_{i=1}^{n}\bar{f}(i,n)\log\left[\bar{f}(i,n)\right]\frac{b-a}{n} = \int_a^b f(x)\log f(x)dx. \tag{C.8}$$

The two limits in Eqs. (C.7) and (C.8) are basically the definition of the Riemann integral. They are presumed to be finite [note, however, that the quantity in Eq. (C.8) might be either positive or negative]. Hence, in this limit, we have

$$H = \lim_{n\to\infty} H(n) = -\int_a^b f(x)\log f(x)dx - \lim_{n\to\infty}\log\left[\frac{b-a}{n}\right]. \tag{C.9}$$

Clearly, the second term on the right hand side of Eq. (C.9) diverges when $n \to \infty$. The reason for this divergence is clear. The larger the n, the larger the number of intervals, and the more information is needed to locate a point on the segment (a, b). Note, however, that this divergent term does not depend on the distribution density $f(x)$. It depends only on how we have divided the segment (a, b). Therefore, when we calculate *differences* in H for different distributions, say $f(x)$ and $g(x)$, we can take the limit $n \to \infty$, *after* the formation of the difference, i.e.

$$\Delta H = \lim_{n\to\infty} \Delta H(n)$$
$$= -\int_a^b f(x)\log f(x)dx - \int_a^b g(x)\log g(x)dx. \qquad \text{(C.10)}$$

Here, the divergent part does not appear and the quantity ΔH is finite in the limit $n \to \infty$. It should also be noted that the quantity H always depends on how accurately we are interested in locating a particle in a segment (a, b) (or in a volume V in the 3 case); in other words, H depends on δ. It is only the difference ΔH that is independent of δ. In practice, we always have a limited accuracy for any measurable quantity, and therefore the strict mathematical limit of $n \to \infty$ is never used in practice.

Thus, for a continuous rv, we will always use, as Shannon did, the definition of the SMI as

$$H = -\int_a^b f(x)\log f(x)dx. \qquad \text{(C.11)}$$

One should be careful, however, in using this definition of the SMI. This SMI does not have all the properties of H as for the finite case. We will use the definition (C.11) for H only when we are interested in the *differences* in the SMI, in which case no difficulties arise even in the limit $n \to \infty$.

For two distribution densities $f(x)$ and $g(x)$ for which the two integrals $-\int_{-\infty}^{\infty} f(x)\log g(x)dx$ and $-\int_{-\infty}^{\infty} f(x)\log f(x)dx$ exist, we have the inequality

$$-\int_{-\infty}^{\infty} f(x)\log\frac{f(x)}{g(x)}\,dx \leq 0. \tag{C.12}$$

The equality holds if and only if $f(x) = g(x)$ for almost all x.

It should be noted that the quantity $I(X;Y)$, the MI for X and Y, is defined as the difference between two uncertainties, thus there exists no problem in extending this quantity to the case of continuous rv's. Hence, for two rv's with densities $f(x)$ and $g(y)$, we define

$$I(X;Y) = \int_{-\infty}^{\infty}\int_{-\infty}^{\infty} f(x,y)\log\left[\frac{F(x,y)}{f(x)g(y)}\right]dxdy, \tag{C.13}$$

where $F(x,y)$ is the density distribution for the joint distribution of X and Y, and $f(x)$ and $g(y)$ are the marginal distributions:

$$f(x) = \int F(x,y)dy, \tag{C.14}$$

$$g(y) = \int F(x,y)dx. \tag{C.15}$$

Appendix D

Functional Derivatives and Functional Taylor Expansion

We present here a simplified definition of the operations of functional derivatives and functional Taylor expansion. It is based on a formal generalization of the corresponding operations applied to functions of a finite number of independent variables.

Consider first a function of one variable:

$$y = f(x). \tag{D.1}$$

The derivative of this function with respect to x is defined as

$$\frac{dy}{dx} = \lim_{\varepsilon \to 0} \frac{f(x+\varepsilon) - f(x)}{\varepsilon}. \tag{D.2}$$

We assume that the function is differentiable, i.e. the limit is that Eq. (D.2) exists.

The derivative itself is a function of x and may be evaluated at any point x_0 in the region where it is defined.

For example,

$$\begin{aligned} y &= f(x) = ax, \\ \frac{dy}{dx} &= \lim_{\varepsilon \to 0} \frac{f(x+\varepsilon) - a(x)}{\varepsilon} = a. \end{aligned} \tag{D.3}$$

Here, the derivative is a constant and has the value a at any point $x = x_0$.

A second example is

$$y = f(x) = x^2. \tag{D.4}$$

The derivative of this function is

$$\frac{dy}{dx} = 2x. \tag{D.5}$$

Clearly, the value of the derivative at the point $x = x_0$ is $2x_0$.

Next, consider a function of n independent variables, $x_1, \ldots, x_n$. A simple example of such a function is

$$f(x_1, \ldots, x_n) = \sum_{i=1}^{n} a_i x_i. \tag{D.6}$$

The partial derivative of f with respect to, say, x_j is

$$\frac{\partial f}{\partial x} = \sum_{i=1}^{n} a_i \frac{\partial x_i}{\partial x_j} = \sum_{i=1}^{n} a_i \delta_{ij} = a_j, \tag{D.7}$$

where δ_{ij} is the Kronecker delta function, defined by

$$\delta_{ij} = \begin{cases} 0 & \text{if } i \neq j, \\ 1 & \text{if } i = j. \end{cases} \tag{D.8}$$

A further generalization of Eq. (D.6) is the case of the vector $\boldsymbol{y} = (y_1, \ldots, y_n)$, which is a function of the vector $\boldsymbol{x} = (x_1, \ldots, x_n)$. This connection can be written symbolically as

$$\boldsymbol{y} = \boldsymbol{F}\boldsymbol{x}, \tag{D.9}$$

where $\boldsymbol{F}$ is a matrix operating on a vector $\boldsymbol{x}$ to produce a new vector, $\boldsymbol{y}$.

We now generalize Eq. (D.9) as follows. We first rewrite the vector x in a new notation:

$$\boldsymbol{x} = (x_1, \ldots, x_n) = [x(1), \ldots, x(n)]. \tag{D.10}$$

Here, we simply write the *component* x_i as $x(i)$, where i is a discrete variable; $i = 1, 2, \ldots, n$. We now let i take any value in a continuous range of real numbers, $a \leq i \leq b$. In this way, we get a vector $\boldsymbol{x}$ with an *infinite* number of components, $x(i)$. The functional relation (D.9) is now reinterpreted as a relation between a *function* $\boldsymbol{x}$ and a *function* $\boldsymbol{y}$.

In Eq. (D.9), $\boldsymbol{x}$ and $\boldsymbol{y}$ are vectors, and F is a matrix operating in a vector space (of finite dimensions). When $\boldsymbol{x}$ and $\boldsymbol{y}$ are functions of a continuous variable, say t, they are viewed as vectors with an infinite number of components. In this case, $\boldsymbol{F}$ is an operator acting in a functional space rather than a vector space. A simple relation between two such functions is

$$y(t) = \int_a^b K(s, t) x(s) ds. \tag{D.11}$$

That is to say, for each *function* x whose components are $x(s)$, we get a new *function* $\mathbf{y}$ whose components are $y(t)$. The function $K(s, t)$ is presumed to be known.

Equation (D.11) can also be written symbolically in the same form as Eq. (D.9), where now $\mathbf{F}$ operates on $\mathbf{x}$ and gives the result $\mathbf{y}$.

In the discrete case, for any two components x_i and x_j of the vector $\boldsymbol{x}$, we have the relationship

$$\frac{\partial x_i}{\partial x_j} = \delta_{ij}. \tag{D.12}$$

Similarly, viewing $x(t)$ and $x(s)$ as two "components" of the function x, we write the analog of Eq. (D.12) as

$$\frac{\delta x(t)}{\delta x(s)} = \delta(t - s). \tag{D.13}$$

Here the Dirac delta function replaces the Kronecker delta function in Eq. (D.12). (See Note 1.)

In Eq. (D.7), the quantity $\partial f/\partial x_J$ is referred to as the *partial derivative* of f with respect to the component x_J. Similarly, the functional derivative of $y(t)$ in Eq. (D.11) with respect to the "component" $x(s')$ is

$$\frac{\delta y(t)}{\delta x(s')} = \int K(s,t)\frac{\delta x(s)}{\delta x(s')}ds = \int K(s,t)\delta(s-s')ds = K(s,t). \tag{D.14}$$

As an example, suppose that we can write the average volume of a system as

$$V = \int_0^\infty \phi N(\phi)d\phi. \tag{D.15}$$

[For an explicit example, see Ben-Naim (2006a).] The functional derivative of V with respect to the component $N(\phi')$ is

$$\frac{\delta V}{\delta N(\phi')} = \int_0^\infty \phi\,\frac{\delta N(\phi)}{\delta N(\phi')}\,d\phi = \int_0^\infty \phi\delta(\phi-\phi')\,d\phi = \phi'. \tag{D.16}$$

Before turning to functional Taylor expansion, we note that many operations with ordinary derivatives can be extended to functional derivatives. We note, in particular, the chain rule of differentiation.

For functions of one variable $y = f(x)$ and $y = g(t)$, we have

$$\frac{dy}{dx}\frac{dx}{dt} = \frac{dy}{dt} \tag{D.17}$$

and, in particular,

$$\frac{dy}{dx}\frac{dx}{dy} = 1. \tag{D.18}$$

In the case of functions of n variables, say

$$y_k = f_k(x_1, \ldots, x_n), \quad k = 1, \ldots, n, \tag{D.19}$$

we have

$$\frac{dy_k}{dt} = \sum_{i=1}^{n} \frac{\partial y_k}{\partial x_i} \frac{dx_i}{dt} \tag{D.20}$$

and, in particular,

$$\delta_{kj} = \frac{dy_k}{dy_j} = \sum_{i=1}^{n} \frac{\partial y_k}{\partial x_i} \frac{\partial x_i}{\partial y_j}. \tag{D.21}$$

The generalization of Eq. (D.21) for the case of functional space is straightforward. We view the equation $\boldsymbol{y} = \boldsymbol{F}\boldsymbol{x}$ as a relationship between the two functions whose components are $y(t)$ and $x(t)$, respectively, and write by analogy with Eq. (D.21)

$$\frac{\delta y(t)}{\delta y(v)} = \int \frac{\delta y(t)}{\delta x(s)} \frac{\delta x(s)}{\delta y(v)} ds = \delta(t-v). \tag{D.22}$$

Here, integration replaces the summation in Eq. (D.21).

We now consider functional Taylor expansion. We start with a simple function of one variable, $f(x)$, for which the Taylor expansion about $x = 0$ is

$$f(x) = f(0) + \left.\frac{\partial f}{\partial x}\right|_{x=0} x + \frac{1}{2} \left.\frac{\partial^2 f}{\partial x^2}\right|_{x=0} x^2 + \cdots. \tag{D.23}$$

As an example, consider $f(x) = a + bx$. Then, we have

$$f(0) = a, \quad \partial f/\partial x = b. \tag{D.24}$$

Hence, from Eq. (D.23), we get

$$f(x) = a + bx, \tag{D.25}$$

i.e. the expansion to first order in x is, in this case, *exact* for any x. In general, if we take the first order expansion

$$f(x) = f(0) + \left.\frac{\partial f}{\partial x}\right|_{x=0} x, \tag{D.26}$$

we get an approximate value for $f(x)$. The quality of the approximation depends on x and on the function f.

For a function of n variables $f(x_1, \ldots, x_n)$, the Taylor expansion about the point $x = 0$ is

$$f(x_1, \ldots, x_n) = f(0, \ldots, 0) + \sum_{i=1}^{n} \left. \frac{\partial f}{\partial x_i} \right|_{x=0} x_i + \cdots , \tag{D.27}$$

where all the derivatives are evaluated at the point $x = 0$.

The generalization to the continuous case is, by analogy with Eq. (D.27),

$$f(x) = f(0) + \int \left. \frac{\delta f(x)}{\delta x(t)} \right|_{x=0} x(t) dt + \cdots . \tag{D.28}$$

Here, the partial derivative has become the functional derivative, and the summation over i has become the integration over t.

As for the nomenclature, the quantity $\partial f / \partial x_i|_{x=0}$ in Eq. (D.27) is referred to as the *partial derivative* of f with respect to the *component* x_i, evaluated at the point $x = 0$. Similarly, in Eq. (D.28), we have the functional derivative of f with respect to the "component" $x(t)$, evaluated at the point $x = 0$.

Appendix E

Some Inequalities for Convex Functions

In this appendix, we will present some important inequalities that are used in connection with IT. Proofs can be found in any textbook of mathematics — in particular, for IT, Yaglom and Yaglom (1983).

Definition

A function $f(x)$ is said to be convex (convex upward or concave downward) in some region (a, b) if for any two points x' and x'' such that $a \le x' \le b, a \le x'' \le b$, the entire straight line connecting the two points $f(x')$ and $f(x'')$ is always *below* the function $f(x)$; see Fig. E.1.

An important property of a convex function is the following: A function $f(x)$ is convex in (a, b) if and only if the second derivative is negative in (a, b).

Examples of convex functions are (we use natural log for convenience)

(1) $$f(x) = \log x. \tag{E.1}$$

This function is defined for $x > 0$. The first two derivatives are

$$f'(x) = \frac{1}{x}, \quad f''(x) = \frac{-1}{x^2} < 0, \quad \text{for } x > 0. \tag{E.2}$$

(2) $$f(x) = -x \log x. \tag{E.3}$$

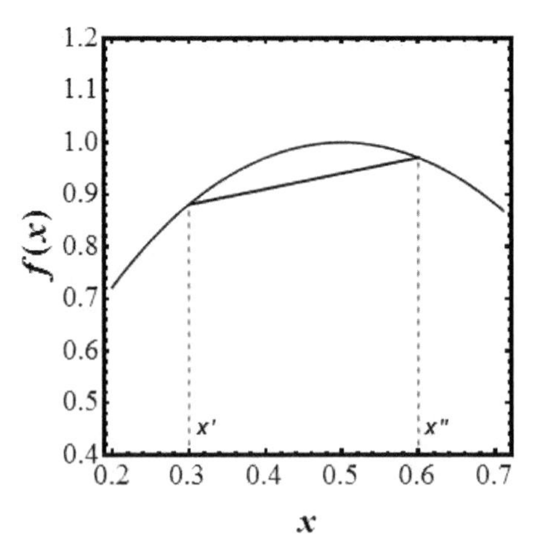

Fig. E.1. A convex function.

This function is defined for $x \geq 0$, and its first two derivatives are

$$f'(x) = -1 - \log x, \quad f''(x) = \frac{-1}{x} < 0, \quad \text{for } x \geq 0. \tag{E.4}$$

(3) The function

$$-x \log x - (1-x)\log(1-x), \quad \text{for } 0 \leq x \leq 1, \tag{E.5}$$

is convex in $0 \leq x \leq 1$, and its first two derivatives are

$$f'(x) = -\log x + \log(1-x), \ f''(x) = \frac{-1}{x(1-x)} < 0, \ \text{for } 0 < x < 1. \tag{E.6}$$

We now present a few inequalities for convex functions:

Inequality I

If $f(x)$ is convex in (a, b), and if $a \leq x_1 \leq x_2 \leq b$, then

$$\frac{f(x_1) + f(x_2)}{2} < f\left(\frac{x_1 + x_2}{2}\right), \tag{E.7}$$

i.e. the value of the function f at the point $\frac{x_1+x_2}{2}$ is always larger than the point $(f(x_1)+f(x_2))/2$ on the straight line connecting the points $f(x_1)$ and $f(x_2)$. This property follows immediately from the definition, and it is clear from Fig. E.1. A more general statement of this property is:

Inequality II

$$\lambda f(x_1) + (1-\lambda)f(x_2) < f(\lambda x_1 + (1-\lambda)x_2), \tag{E.8}$$

for any λ, such that $0 \leq \lambda \leq 1$.

As an example, we begin by applying the inequality (E.7) to the function (E.1). We get

$$\frac{\log x_1 + \log x_2}{2} < \log\left(\frac{x_1+x_2}{2}\right) \tag{E.9}$$

or, equivalently,

$$(x_1 x_2)^{\frac{1}{2}} < \frac{x_1+x_2}{2}. \tag{E.10}$$

This is a well-known inequality, i.e. the geometric average of two numbers x_1, x_2 is always smaller than the arithmetic average.

Applying inequality II to this function, we obtain

$$\lambda \log x_1 + (1-\lambda)\log x_2 < \log(\lambda x_1 + (1-\lambda)x_2), \quad \text{for } 0 \leq \lambda \leq 1, \tag{E.11}$$

or, equivalently,

$$x_1^{\lambda} x_2^{(1-\lambda)} < \lambda x_1 + (1-\lambda)x_2. \tag{E.12}$$

Inequality III

Another generalization of inequality I is

$$\frac{f(x_1) + f(x_2) + \cdots + f(x_m)}{m} < f\left(\frac{x_1 + x_2 + \cdots + x_m}{m}\right) \tag{E.13}$$

or, more generally:

Inequality IV

$$\lambda_1 f(x_1) + \lambda_2 f(x_2) + \cdots + \lambda_m f(x_m) < f(\lambda_1 x_1 + \lambda_2 x_2 + \cdots + \lambda_m x_m), \tag{E.14}$$

for $\lambda_i \geq 0$ and $\sum_{i=1}^{m} \lambda_i = 1$.

Applying inequality IV for the function (E.1), we have

$$\sum_{i=1}^{m} \lambda_i \log x_i \leq \log\left(\sum_{i=1}^{m} \lambda_i x_i\right) \tag{E.15}$$

or, equivalently,

$$\prod_{i=1}^{m} x_i^{\lambda_i} \leq \sum_{i=1}^{m} \lambda_i x_i. \tag{E.16}$$

An important inequality that follows from (E.15) is the following:

Let $p_1, \ldots, p_m$ and $q_1, \ldots, q_m$ be any of the two distributions, such that $\sum_{i=1}^{m} p_i = \sum_{i=1}^{m} q_i = 1$, for $p_i > 0$ and $q_i > 0$. We choose $\lambda_i = p_i$ and $x_i = \frac{q_i}{p_i}$. Since $\sum \lambda_i = 1$, and since the function $\log x$ is convex for any $x > 0$, application of (E.15) for this choice gives

$$\sum_{i=1}^{m} p_i \log \frac{q_i}{p_i} \leq \log\left(\sum_{i=1}^{m} p_i \frac{q_i}{p_i}\right) = \log\left(\sum_{i=1}^{m} p_i\right) = 0. \tag{E.17}$$

Hence, it follows from (E.17) that

$$\sum_{i=1}^{m} p_i \log q_i \leq \sum_{i=1}^{m} p_i \log p_i. \tag{E.18}$$

This inequality can be proved from the inequality $\log x \leq x - 1$ for any $x > 0$. The latter inequality also follows from the convexity

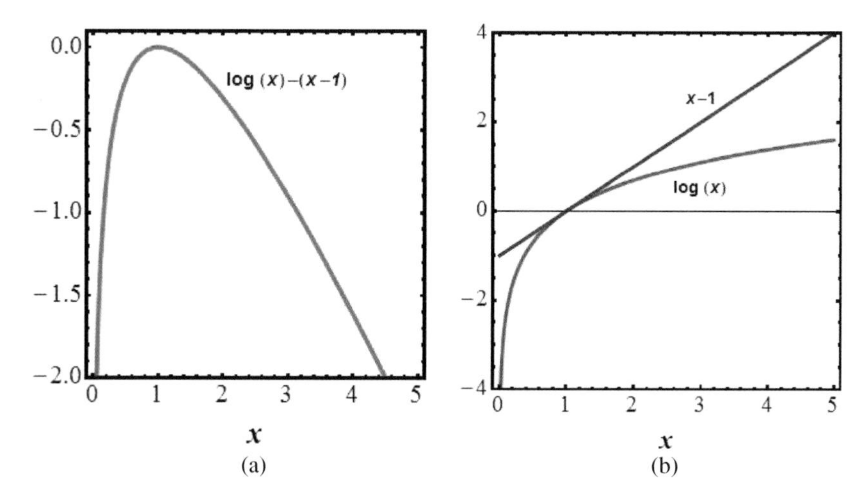

Fig. E.2. (a) The function $\log(x) - (x - 1)$ and (b) the functions $\log(x)$ and $(x - 1)$.

of the function $f(x) = \log x - (x - 1)$, the derivative of which is

$$f'(x) = \frac{1}{x} - 1 = \frac{1 - x}{x}. \tag{E.19}$$

For $0 < x < 1$, this derivative is positive. Hence, in this region, $f(x)$ is a monotonically increasing function of x. Since $f(1) = 0$, it follows that $f(x) \leq 0$, or, equivalently, $\ln x \leq x - 1$ for $x < 1$.

For $x > 1$, $f'(x)$ is negative, and hence $f(x)$ is a monotonically decreasing function of x. Since $f(1) = 0$, it follows that $f(x) \leq 0$ or, equivalently, $\log x \leq x - 1$, for $x > 1$. The equality sign holds for $x = 1$. Figure E.2 shows the function $\log x - (x - 1)$ and the functions $\log x$ and $x - 1$, separately. The geometric meaning of the inequality $\log x \leq x - 1$ is that the straight line $y = x - 1$ is always above the line $y = \log x$, for any $x > 0$.

Another application of (E.14), for the function (E.3), is

$$-\sum_{i=1}^{m} \lambda_i x_i \log x_i < -\left(\sum_{i=1}^{m} \lambda_i x_i\right) \log\left(\sum_{i=1}^{m} \lambda_i x_i\right), \tag{E.20}$$

for $\lambda_i \geq 0, \quad \sum \lambda_i = 1$.

An important conclusion results from the convexity of the function $-x \log x$.

The SMI of any experiment having m outcomes cannot exceed the SMI of an experiment with m equally probable outcomes. Thus, if we choose $x_i = p_i$, the probabilities of the outcomes of an experiment we have from (E.13),

$$-\frac{\sum x_i \log x_i}{m} < -\left(\frac{\sum x_i}{m}\right) \log \left(\frac{\sum x_i}{m}\right), \tag{E.21}$$

multiplying by m and choosing $x_i = p_i$, $\sum x_i = 1$, we get

$$-\sum_{i=1}^{m} p_i \log p_i \leq -\log \frac{1}{m} = \log m; \tag{E.22}$$

i.e. the SMI of an experiment with m equally probable events is always larger than the SMI of m events with unequal probability. The equality holds if and if only all $p_i = p$.

Appendix F

Distribution Functions in 1D Models

There is a standard method of calculating the thermodynamic properties of a 1D model system. We will present here only an outline of the general method, and then we derive the various probability distributions. A more complete treatment can be found in Ben-Naim (1992, 2001, 2014).

The model

We consider a system of N units on linear lattice points in one dimension. Each unit can be in one of two states: up or down for spins, an occupied or empty site in a lattice gas, a site occupied by molecule A or B, etc.

The *configuration* of the entire system is given by the N-dimensional vector $\mathbf{s} = (s_1, s_2, \ldots, s_N)$. Thus, spins s_i can either be up (+1), or down (−1). We assume that the total energy of the system is given by

$$E(\mathbf{s}) = -J \sum_{i=1}^{N-1} s_i s_{i+1} - H \sum_{i=1}^{N} s_i. \tag{F.1}$$

The first term on the right hand side of Eq. (F.1) is the sum of all interaction energies between successive spins (here we assume only nearest neighbors interaction between spins at locations i and $i+1$). In addition, there is an interaction energy between each spin and an external field H (not to be confused with SMI, which is also denoted by H).

The probability of observing a specific configuration $\mathbf{s} = (s_1, \ldots, s_N)$ is given by

$$P(\mathbf{s}) = \frac{\exp[-E(\mathbf{s})/k_B T]}{Q}, \tag{F.2}$$

where k_B is Boltzmann's constant, T the absolute temperature, and Q the normalization constant referred to as the partition function. Henceforth, we will take $k_B = 1$ so that T is measured in units of energy:

$$Q = \sum_{s} \exp\left[\frac{-E(\mathbf{s})}{T}\right]. \tag{F.3}$$

The partition function defined in Eq. (F.3) is used to calculate all the thermodynamic quantities of the system. In this appendix, we will be interested only in the various probability distributions. We will refer to $\mathbf{s}$ as the microscopic state of the entire system, and s_i is the state of a single spin (or a single site).

It is convenient to *close* the chain of particles in such a way that particle $i = 1$ interacts with particle $i = N$. This leads to a simple expression for the partition function Q:

$$Q = \mathrm{Tr}(P^N), \tag{F.4}$$

where P is the 2×2 matrix defined by

$$P = \begin{pmatrix} \exp(J + H/T) & \exp(-J/T) \\ \exp(-J/T) & \exp(J - H/T) \end{pmatrix}. \tag{F.5}$$

In Eq. (F.4), the partition function is given as the *trace* of the Nth power of the matrix P. This is very convenient, since we can express the trace of the matrix P^N in terms of the eigenvalues of the matrix P. The latter are obtained from solving the secular equation

$$|P - \lambda \mathbf{I}| = 0. \tag{F.6}$$

The two solutions to Eq. (F.6) are

$$\lambda_{\pm} = \exp\left(\frac{J}{T}\right)\left[\cosh(x) \pm \sqrt{\sin h^2(x) + \exp(-4J/T)}\right], \tag{F.7}$$

where $x = \exp(H/T)$, and sinh and cosh are the hyperbolic sine and cosine, defined by

$$\sinh(x) = \frac{e^x - e^{-x}}{2},$$
$$\cosh(x) = \frac{e^x + e^{-x}}{2}. \tag{F.8}$$

Thus, we can have the eigenvalues λ_+ and λ_-, and we can write explicitly the partition function:

$$Q = \mathrm{Tr}(P^N) = \lambda_+^N + \lambda_-^N. \tag{F.9}$$

Therefore, solving the secular question (F.6) provides us with the eigenvalues λ_+ and λ_-, which in turn provide us with an explicit dependence of the partition function in terms of the molecular parameters of the system.

In statistical mechanics, one usually takes the limit of Eq. (F.9) for very large $N, N \to \infty$, in which case one can use only the largest eigenvalue, in our case λ_+. However, for calculating the various distribution functions we will need both of the eigenvalues λ_+ and λ_-, as well as the eigenvectors. The eigenvector of the matrix P corresponding to the eigenvalue λ_i is defined as the column vector $|a_i\rangle$, which fulfills the equation

$$P|a_i\rangle = \lambda_i |a_i\rangle. \tag{F.10}$$

The single distribution function

Once we have an explicit expression for Q, we can use the probability distribution in Eq. (F.2) to calculate any molecular distribution functions. These are essentially the marginal probabilities of $P(s)$. For instance, the singlet molecular distribution function is defined by

$$P(s_1 = \alpha) = \sum_{s_2,\ldots,s_N} P(s_1, s_2, \ldots, s_N). \tag{F.11}$$

This is the probability of finding a specific spin in a state α. Note that since all the spins are equivalent, the singlet probability does not depend on which specific spin we choose; here we choose the one with index 1. On the right hand side of Eq. (F.11), we sum over all the states of the $N - 1$ spins (each s_i can be either 1 or -1, corresponding to "up" or "down," respectively). This gives the singlet distribution as a marginal probability of $P(\mathbf{s})$.

We next express $P(s_1 = \alpha)$ in terms of the parameters of the model. We rewrite Eq. (F.11) as

$$\begin{aligned} P(s_1 = \alpha) &= \sum_{s_2,\ldots,s_N} P(\alpha, s_2, \ldots, s_N) \\ &= \sum_{s_1,s_2,\ldots,s_N} P(s_1, s_2, \ldots, s_N)\delta_{s_1,\alpha}, \end{aligned} \tag{F.12}$$

where $\delta_{s_1,\alpha}$ is the Kronecker delta function, i.e.

$$\delta_{s_1,\alpha} = \left\{ \begin{matrix} 1 & \text{if} & s_1 = \alpha \\ 0 & \text{if} & s_1 \neq \alpha \end{matrix} \right\}. \tag{F.13}$$

Note that in the last term on the right hand side of Eq. (F.12) we added the delta function and summed over *all* the states of the N spins.

Using similar steps which lead us from the definition of Q in Eq. (F.3) to its final form Eq. (F.9), we can also express the sum on the right hand side of Eq. (F.12) in terms of the elements of the matrix P, i.e.

$$P(s_1 = 1) = \langle 1|P^N|1\rangle,$$
$$P(s_1 = -1) = \langle -1|P^N| - 1\rangle, \tag{F.14}$$

where $\langle 1|$ and $|1\rangle$ are the unit row and column vector (1, 0). Similarly, $\langle -1|$ and $|-1\rangle$ are the unit row and column vector (0, 1), respectively.

Denoting by $|a_i\rangle$ and by $\langle a_i|$ the column and the row eigenvectors corresponding to the eigenvalue λ_i, we can use the identity for the unit matrix $\boldsymbol{I}$ in terms of these eigenvectors:

$$\boldsymbol{I} = \sum_i |a_i\rangle\langle a_i|, \tag{F.15}$$

where on the right hand side we multiply a column vector $|a_i\rangle$ by a row vector $\langle a_i|$, and sum over all possible states of a single spin (here only two states, 1 and -1, corresponding to "up" and "down," respectively). With the identity (F.15), and some simple algebra [for details, see Ben-Naim (1992, 2014)], we can rewrite the singlet distribution as

$$P(s_1 = 1) = \langle 1|a_1\rangle^2 = a_{11}^2, \tag{F.16}$$

$$P(s_1 = 1) = \langle -1|a_1\rangle^2 = a_{12}^2, \tag{F.17}$$

where a_{11} and a_{12} are the first and second components of the eigenvector a_1. We also have the normalization condition:

$$P(s_1 = 1) + P(s_1 = -1) = a_{11}^2 + a_{12}^2 = 1. \tag{F.18}$$

The pair distribution function

The pair distribution function is defined in analogy with Eq. (F.12) as

$$P(s_1 = \alpha; s_j = \gamma) = \sum_{\substack{\text{overall } s_i \\ \text{except } s_1 \text{ and } s_j}} P(s_1 = \alpha, s_2, \ldots, s_j = \gamma, \ldots, s_N)$$

$$= \sum_{\text{all } s_i} P(s_1, s_2, \ldots, s_N)\delta_{s_1,\alpha}\delta_{s_j,\gamma}. \tag{F.19}$$

This is the probability of finding spin 1 in state α (α can be either 1 or -1, corresponding to "up" or "down"), and spin j in state γ (again, γ can be either 1 or -1). Note again that the pair distribution depends only on the *distance* $j-1$, between the spins, and not on the particular locations of the two spins.

After a lengthy algebra, one can get the results for all the four pair distribution functions. For instance, the probability of finding one spin "up," and $j-1$ units apart from a second spin in the state "down," is

$$P(s_1 = \text{up}; s_j = \text{down}) = a_{11}^2 a_{12}^2 + \left(\frac{\lambda_-}{\lambda_+}\right)^{j-1} a_{11}a_{12}a_{21}a_{22}. \tag{F.20}$$

Note that a_{11} and a_{12} are the first and second components of the eigenvector a_1. Similarly, a_{21} and a_{22} are the first and second components of the eigenvector a_2, while λ_- and λ_+ are the two eigenvalues with $\lambda_+ > \lambda_-$.

Thus, by finding the eigenvalues and the eigenvectors of the matrix P, we can calculate all the pair distribution functions; $j = 2$ corresponds to nearest neighbors, $j = 3$ corresponds to next nearest neighbors, etc.

One can also calculate the triplet and higher order distribution functions for this model. [See Ben-Naim (1992).] Here, we note only that the conditional probabilities all have the Markovian property.

For instance,

$$\begin{aligned} P(s_j = \alpha | s_{j-1} = \gamma, s_{j-2} = \gamma, \ldots, s_1 = \gamma) \\ = P(s_j = \alpha | s_{j-1} = \gamma) \\ = P(s_2 = \alpha | s_1 = \gamma). \end{aligned} \tag{F.21}$$

This means that the probability of finding s_j in state α, given the states of all the $j - 1$ preceding spins, depends only on the state of the nearest preceding spin.

Appendix G

Entropy Change in an Expansion Process in a Gravitational Field

In this appendix, we briefly discuss the method for calculating the entropy of an ideal gas (no intermolecular interaction) in a gravitational field. A more detailed discussion can be found in Ben-Naim (1992). The reader who is not familiar with statistical mechanics may skip this appendix.

We start with the partition function of N simple particles, in volume V and at temperature T, in a gravitational field in the direction of the z axis:

$$Q(T, V, N; g) = \frac{1}{N!\Lambda^{3N}} \int_V \exp[-\beta U(\mathbf{R}_1, \ldots, \mathbf{R}_N)] d\mathbf{R}_1 \cdots d\mathbf{R}_N, \tag{G.1}$$

with Λ^3 being the momentum partition function,

$$\Lambda = \frac{h^2}{\sqrt{2\pi m k_B T}}, \tag{G.2}$$

where h is the Planck constant, m the mass of the particles, k_B the Boltzmann constant, and $\beta = 1/k_B T$. The potential energy is given by

$$U(\mathbf{R}_1, \ldots, \mathbf{R}_N) = \sum_{i=1}^{N} -mgz_i, \tag{G.3}$$

where g is the gravitational acceleration and z_i is the height of the ith particle.

If the system has a cross-section of area a, and each z_i can vary between the heights of $h_1 \leq z_i \leq h_2$, we can rewrite the integral in Eq. (G.1) as

$$\int \exp[-\beta U(\mathbf{R}_1, \ldots, \mathbf{R}_N)] d\mathbf{R}_1 \cdots d\mathbf{R}_N = a^N \left\{ \int_{h1}^{h2} \exp(-\beta mgz) dz^N \right\}. \tag{G.4}$$

Hence, the partition function is

$$Q(T, V, N; g) = \frac{V^N}{N! \Lambda^{3N}} q_g^N, \tag{G.5}$$

where

$$q_g = \frac{1}{h_2 - h_1} \int_{h_1}^{h_2} \exp(-\beta mgz) dz. \tag{G.6}$$

The Helmholtz energy of the system is

$$A(T, V, N; g) = -k_B T \ln Q(T, V, N : g) \tag{G.7}$$

and the chemical potential is

$$\begin{aligned} \mu = \left(\frac{\partial A}{\partial N} \right)_{T,V} &= RT \ln \frac{N}{V} \Lambda^3 q_g^{-1} \\ &= \mu(g = 0) + RT \ln q_g^{-1}. \end{aligned} \tag{G.8}$$

In Eq. (G.8), we have used $R = k_B N_{\text{av}}$ instead of k_B; $N_{\text{av}} \approx 6 \times 10^{23}$ is the Avogadro number. The first term on the right hand side of the equation is the chemical potential of the same system in the absence of the gravitational field ($g = 0$), and the second term is the addition to the chemical potential upon "turning on" the gravitational field.

The entropy per mole is calculated from

$$s = -\left(\frac{\partial \mu}{\partial T}\right)_V = s(g=0) - \frac{\partial}{\partial T}(RT\ln q_g)$$
$$= s(g=0) + R\ln q_g + N_{av}\frac{mg\bar{z}}{T}. \quad \text{(G.9)}$$

In Eq. (G.9), $s(g = 0)$ is the entropy of the same system in the absence of the field, and $mg\bar{z}$ is the average potential energy of the particles, defined by

$$mg\bar{z} = \int_{h_1}^{h_2} mgz\, P(z)dz, \quad \text{(G.10)}$$

where $P(z)$ is the probability density of finding a particle between z and $z + dz$, and is given by

$$P(z) = \frac{\exp(-\beta mgz)}{\int_{h_1}^{h_2} \exp(-\beta mgz)dz}. \quad \text{(G.11)}$$

Using this probability density, we can rewrite the entropy in Eq. (G.9) as

$$s = s(g=0) - R\int P(z)\ln P(z)dz - R\ln(h_2 - h_1). \quad \text{(G.12)}$$

We see that the entropy of the system has two parts: the entropy of the same system in the absence of the field $s(g = 0)$, plus the change in the entropy due to the redistribution of the particles along the z axis; from the uniform distribution to the final distribution $P(z)$. Thus, the sum of the last two terms on the right hand side of Eq. (G.12) is the difference in the locational SMI along the z axis, which occurs upon "turning on" the gravitational field.

Notes

Notes on Chapter 1

Note 1

There have been a few attempts to construct a semantic theory of information, which we will not discuss in this book.

Note 2

In this paragraph, Shannon makes two errors. The first is that he chooses to call his measure "entropy." As we will see, this has caused a maelstrom in both IT and thermodynamics. Second, the famous *H* in Boltzmann's *H*-theorem is an SMI but not entropy. The identification of the *H*-function in Boltzmann's *H*-theorem with entropy is at the heart of the so-called reversibility paradox. This paradox is discussed in Ben-Naim (2015a, 2016a).

Notes on Chapter 2

Note 1 (answers to Exercise 2.1)

Note that since you do not know the probabilities as provided in Fig. 2.2, the answers given in cases $a - f$ should be the same.

(1) Since you do not know the probabilities, you have total uncertainty (in a colloquial sense) about where the coin is.
(2) The lacking information is the answer to the question "In which box is the coin?"
(3) Qualitatively speaking, the *amount* of information you lacked is the same for the three cases. Since the probabilities are not known, the SMI is not known either.

Note 2 (answers to Exercise 2.2)

As in Exercise 2.1, since you do not know the probabilities, the answers you give to questions 1–4 are the same for cases $a - f$.

(1) One cannot measure the extent of the surprise upon being given the information about the location of the coin.
(2) Before being informed, you had *total* uncertainty about the location of the coin; this uncertainty was removed upon hearing that the coin is in box A. The answer is the same for all of the cases $a - f$.
(3) The *information* you got is "The coin is in box A." This information is the same for cases $a - f$.
(4) The amount of (colloquial) information is the same in cases $a - f$. The amount depends on how you measure the information contained in the sentence "The coin is in box A." You can measure the amount of information according to the number of words, the length of the sentence, etc. All these are different from the SMI associated with the probabilities given in Fig. 2.2, but of which you are not aware.

Note 3 (answers to Exercise 2.3)

Note carefully that in this case you *know* the probabilities as shown in Fig. 2.2. Therefore, your answers should be different for the cases $a - f$.

(1) Now that you know the probabilities as provided in Fig. 2.2, your level of surprise at finding that the coin is in box A is different for each case.

(a) You know for certain that the coin is in A (with probability 1). Therefore, you are not surprised when you are informed that the coin is in box A. You can say that your surprise is zero in this case.
(b) You know that the probability of finding the coin in box A is high, and therefore your level of surprise is low.
(c) You know that the probability of finding the coin in box A is 2/3, i.e. twice as large as in box B. Therefore, you are not very surprised, or your level of surprise is low.
(d) In this case, you know that there is no preference for either box A or box B, and therefore you are not very surprised, or you are mildly surprised by finding out that the coin is in box A.
(e) In this case, you know that the probability of finding the coin in box A is very small, and therefore you should be quite surprised to learn that it was found in box A. Thus, the smaller the probability p_A, say 1/100, 1/1000, 1/10,000, the higher the level of your surprise.
(f) In this case, you know that the probability of finding the coin in box A is nil. This is tantamount to being *certain* that the coin is in box B. When you hear that the coin was found in box A, you can say that you are infinitely surprised. "Infinitely surprised" is a way of extrapolating from a series of probabilities:

$$p_A = \frac{1}{100}, \frac{1}{1000}, \frac{1}{10,000} \to 0.$$

The smaller the probability, the higher the level of your surprise. Therefore, extrapolating to $p_A \to 0$ will lead to infinite surprise. However, since you were *certain* that the coin was not in box A (but in box B), *"infinitely surprised"* can also be translated into not believing the news about the coin's location.

(2) As in the answer to question 1 given above, you can say that the larger the uncertainty about the location of the coin, the smaller the probability of finding the coin in box A. Therefore, the extent of the removal of uncertainty is smallest (or nil) in the case of a, and it increases from a to b to c to d to e. In case f, you are certain that the coin is not in A. Therefore, it would be awkward to say that the removal of uncertainty is largest in this case. Thus, if you knew for certain that the coin was in box B, and you are told that it was found in box A, you should not believe this information, although by extrapolation from a series of decreasing p_A you can conclude that the removal of uncertainty is infinity.

(3) The information you got when you were informed that the coin was found in box A is "The coin is in box A." This information (colloquially speaking) is the same for cases $a - f$.

(4) How much information you got depends on how you measure the information you received. If you measure the information in terms of the number of letters or number of words in the message "The coin is in box A," then it is the same for all of the cases $a - f$. You can also assign an SMI to the entire message, "The coin is in box A," taking into account the frequencies of occurrence of the various letters in the English language. In this case, the SMI associated with this message is the same for each of the cases $a - f$ (see also Secs. 2.12 and 2.13 for a definition of the SMI per letter of the English language). However, if you are interested in the SMI associated with the distribution given for cases $a - f$ in Fig. 2.2, then the SMI is not defined for a single event but for the *entire* experiment, i.e. the entire probability distribution. See the next answers to questions 5 and 6.

(5) Here, we asked about the uncertainty about the location of the coin given the probability distribution. Clearly, the uncertainty about the location of the coin is nil for cases a and f. In both

cases, you are certain about the location of the coin. In cases b and e, you are not absolutely certain, but your uncertainty about the location of the coin is not very large. In case c, your uncertainty is larger than in case b (or in case e). Your maximum uncertainty is in case d. Compare your answers to these questions with the values of the SMI defined for each of the cases in Fig. 2.2.

(6) The answer to this question is very similar to the one given to question 5. The values of the SMI for cases $a-f$ are given in Fig. 2.2. The lack of information is minimum for cases a and f and maximum for case d (see Fig. 2.1). Note carefully that the answers to questions 5 and 6 pertain to the *entire* experiment, i.e. to the entire probability distributions, and not to single events.

Note 4 (solution to Exercise 2.4)

The total SMI is

$$H = 8\frac{1}{8}\log_2 8 = 3\,\text{bits}.$$

The "smartest" method

Dividing each time into two equally probable parts. [For details, see Ben-Naim (2008)].

The average gain of information in each step is

$$\begin{aligned}
q_1 &= H\left(\frac{1}{2},\frac{1}{8}\right) = 1\,\text{bit},\\
q_2 &= \frac{1}{2}H\left(\frac{1}{2},\frac{1}{2}\right) + \frac{1}{2}H\left(\frac{1}{2},\frac{1}{2}\right) = 1\,\text{bit},\\
q_3 &= \frac{1}{2}\left[\frac{1}{2}H\left(\frac{1}{2},\frac{1}{2}\right) + \frac{1}{2}H\left(\frac{1}{2},\frac{1}{2}\right)\right]\\
&\quad + \frac{1}{2}\left[\frac{1}{2}H\left(\frac{1}{2},\frac{1}{2}\right) + \frac{1}{4}H\left(\frac{1}{2},\frac{1}{2}\right)\right] = 1\,\text{bit}.
\end{aligned}$$

Thus, in each step, the reduction in the SMI is 1 bit, and hence the total reduction in the SMI is

$$q_1 + q_2 + q_3 = 3.$$

The corresponding probabilities are:

The probability of terminating the game in the first step is 0:

$$P(G_1) = 0,$$
$$P(N_1) = 1.$$

The probability of terminating the game in the second step is also 0:

$$P(G_2, N_1) = P(G_2|N_1)\,P(N_1) = 0 \times 1 = 0,$$
$$P(N_2, N_1) = P(N_2|N_1)\,P(N_1) = 1.$$

The probability of terminating the game in the third step is 1:

$$P(G_3, N_1, N_2) = P(G_3|N_1\,N_2)P(N_1\,, N_2) = 1 \times 1 = 1.$$

The average number of steps is thus

$$0 \times 1 + 0 \times 2 + 1 \times 3 = 3,$$

which is exactly equal to H.

The "dumbest" method

Choosing one box at each step [see Fig. 2.7 and, for more details, Ben-Naim (2008)].

The total SMI is the same; however, the average gain of information in each step is different now:

$$q_1 = H\left(\frac{1}{8}, \frac{7}{8}\right) = 0.543 \text{ bits},$$

$$q_2 = \frac{7}{8} H\left(\frac{1}{7}, \frac{6}{7}\right) = \frac{7}{8} \times 0.592 = 0.518 \text{ bits},$$

$$q_3 = \frac{7}{8} \times \frac{6}{7} H\left(\frac{1}{6}, \frac{5}{6}\right) = \frac{6}{8} \times 0.650 = 0.487 \text{ bits},$$

$$q_4 = \frac{7}{8} \times \frac{6}{7} \times \frac{5}{6} H\left(\frac{1}{5}, \frac{4}{5}\right) = \frac{5}{8} \times 0.722 = 0.451 \text{ bits},$$

$$q_5 = \frac{7}{8} \times \frac{6}{7} \times \frac{5}{6} \times \frac{4}{5} H\left(\frac{1}{4}, \frac{3}{4}\right) = \frac{4}{8} \times 0.8113 = 0.406 \text{ bits}$$

$$q_6 = \frac{7}{8} \times \frac{6}{7} \times \frac{5}{6} \times \frac{4}{5} \times \frac{3}{4} H\left(\frac{1}{3}, \frac{2}{3}\right) = \frac{3}{8} \times 0.918 = 0.344 \text{ bits,}$$

$$q_7 = \frac{7}{8} \times \frac{6}{7} \times \frac{5}{6} \times \frac{4}{5} \times \frac{3}{4} \times \frac{2}{3} H\left(\frac{1}{2}, \frac{1}{2}\right) = \frac{2}{8} \times 1 = 0.25 \text{ bits.}$$

The sum of all average gains of information is

$$\sum_{i=1}^{7} q_i = 3 \text{ bits.}$$

Note again that in this case, in contrast to the previous method, we *can* find the coin in the first step (or the second, the third, etc.); q_i is the average gain of information in the ith step, provided that we reach the ith step.

Note also that the SMI left after each step increases with each step,

$$H\left(\frac{1}{8}, \frac{7}{8}\right) < H\left(\frac{1}{7}, \frac{6}{7}\right) < H\left(\frac{1}{6}, \frac{5}{6}\right) < \cdots H\left(\frac{1}{2}, \frac{1}{2}\right) = 1,$$

but the average gain of information decreases with each step, i.e.

$$q_1 > q_2 > q_3 \cdots q_7 = 0.25 \text{ bits.}$$

The sum of the average gains is

$$\sum_{i=1}^{7} q_1 = 3 \text{ bits.}$$

This is the same as in the smartest method, and also equal to the value of H for this game.

In the smartest method, we gain 1 bit from each question, and hence we need on the average a smaller number of questions. In the dumbest method, less information is acquired on the average in each step. Hence, more questions are needed on the average.

The corresponding probabilities of terminating the gain in each step are

$$
\begin{aligned}
P(G_1) &= \frac{1}{8}, \\
P(N_1) &= \frac{7}{8}, \\
P(G_2, N_1) &= P(G_2|N_1)P(N_1) \\
&= \frac{1}{7} \times \frac{7}{8} = \frac{1}{8}, \\
P(N_2, N_1) &= P(N_2|N_1)P(N_1) \\
&= \frac{6}{7} \times \frac{7}{8} = \frac{6}{8}, \\
P(G_3, N_1, N_2) &= P(G_3|N_1, N_2)P(N_1, N_2) \\
&= \frac{1}{6} \times \frac{6}{8} = \frac{1}{8}, \\
P(N_3, N_1, N_2) &= P(N_3|N_1, N_2)P(N_1, N_2) \\
&= \frac{5}{6} \times \frac{6}{8} = \frac{5}{8}, \\
P(G_4, N_1, N_2, N_3) &= P(G_4|N_1, N_2, N_3)P(N_1, N_2, N_3) \\
&= \frac{1}{5} \times \frac{5}{8} = \frac{1}{8}, \\
P(N_4, N_1, N_2, N_3) &= P(N_4|N_1, N_2, N_3)P(N_1, N_2, N_3) \\
&= \frac{4}{5} \times \frac{5}{8} = \frac{4}{8}, \\
P(G_5, N_1, N_2, N_3, N_4) &= P(G_5|N_1, N_2, N_3, N_4) \\
&\quad \times P(N_1, N_2, N_3, N_4) \\
&= \frac{1}{4} \times \frac{4}{8} = \frac{1}{8}, \\
P(N_5, N_1, N_2, N_3, N_4) &= P(N_5|N_1, N_2, N_3, N_4) \\
&\quad \times P(N_1, N_2, N_3, N_4)
\end{aligned}
$$

$$
\begin{aligned}
&= \frac{3}{4} \times \frac{4}{8} = \frac{3}{8}, \\
P(G_6, N_1, N_2, N_3, N_4, N_5) &= P(G_6 | N_1, N_2, N_3, N_4, N_5) \\
&\quad \times P(N_1, N_2, N_3, N_4, N_5) \\
&= \frac{1}{3} \times \frac{3}{8} = \frac{1}{8}, \\
P(N_6, N_1, N_2, N_3, N_4, N_5) &= P(N_6 | N_1, N_2, N_3, N_4, N_5) \\
&\quad \times P(N_1, N_2, N_3, N_4, N_5) \\
&= \frac{2}{3} \times \frac{3}{8} = \frac{2}{8}, \\
P(G_7, N_1, N_2, N_3, N_4, N_5, N_6) &= P(G_7 | N_1, N_2, N_3, N_4, N_5, N_6) \\
&\quad \times P(N_1, N_2, N_3, N_4, N_5, N_6) \\
&= 1 \times \frac{2}{8} = \frac{2}{8}.
\end{aligned}
$$

The average number of steps in the method is

$$
\begin{aligned}
&\frac{1}{8} \times 1 + \frac{1}{8} \times 2 + \frac{1}{8} \times 3 + \frac{1}{8} \times 4 + \frac{1}{8} \times 5 + \frac{1}{8} \times 6 + \frac{2}{8} \times 7 \\
&\quad = \frac{35}{8} = 4\frac{3}{8}.
\end{aligned}
$$

Note carefully that the average number of steps in this method $\left(4\frac{3}{8}\right)$ is *larger* than the number of steps in the smartest method (3). However, the total amount of information gained is the same (3 bits), independently of the strategy for asking questions.

Note 5 (solution to Exercise 2.5)

The SMI for this game is

$$
H = -\left(\frac{2}{4}\log_2\frac{1}{4} + \frac{1}{2}\log_2\frac{1}{2}\right) = \frac{3}{2} = 1\frac{1}{2} \text{ bits.}
$$

The "smartest" method

Divide each time into two equally probable halves [see diagram in Fig. 2.10(a)]:

$$q_1 = H\left(\frac{1}{2}, \frac{1}{2}\right) = 1,$$

$$q_2 = \frac{1}{2} H\left(\frac{1}{2}, \frac{1}{2}\right) = \frac{1}{2}.$$

The sum of these two steps is

$$q_1 + q_2 = 1\frac{1}{2}.$$

This is equal to H, as we calculated above.

The probabilities of terminating the game in each step are

$$P(G_1) = \frac{1}{2},$$

$$P(N_1) = \frac{1}{2},$$

$$P(G_2, N_1) = P(G_2|N_1)\,P(N_1) = 1 \times \frac{1}{2} = \frac{1}{2}.$$

The average number of steps is

$$\frac{1}{2} \times 1 + \frac{1}{2} \times 2 = \frac{3}{2} = 1\frac{1}{2},$$

which is the same as H as calculated above.

The "dumbest" method [see diagram in Fig. 2.10(b)]

The average gain of information in each step is

$$q_1 = H\left(\frac{1}{4}, \frac{3}{4}\right) = 0.8113 \text{ bits},$$

$$q_2 = \frac{3}{4} H\left(\frac{1}{3}, \frac{2}{3}\right) = 0.6887 \text{ bits}.$$

The sum of these two is

$$q_1 + q_2 = 1.5 \text{ bits},$$

which is the same as H for this case. However, the average number of steps is different.

The probabilities in this case are

$$P(G_1) = \frac{1}{4},$$

$$P(N_1) = \frac{3}{4},$$

$$P(G_2, N_1) = P(G_2|N_1)\,P(N_1) = 1 \times \frac{3}{4}.$$

The average number of steps is thus

$$\frac{1}{4} \times 1 + \frac{3}{4} \times 2 = \frac{7}{4} = 1\frac{3}{4},$$

which is slightly higher than in the "smartest" method.

Note 6 (solution to Exercise 2.6)

For the three cases, the number of questions we need to ask using the smartest strategy is

$$H(8) = \log_2 8 = 3 \text{ bits}, \quad H(16) = 4 \text{ bits}, \quad H(32) = 5 \text{ bits}.$$

By choosing the smartest strategy, we obtain from each question the maximum information, which is 1 bit. Hence, the amount of information is *equal* to the number of questions we need to ask in order to obtain the required information. Using any other strategy for asking questions will be less "efficient," in the sense that each answer provides less than 1 bit of information, and hence we need to ask more questions so as to acquire the same information. Thus, if you pay, say, 1 cent for each answer, then on the average you will pay the least to get the same information when choosing the smartest strategy. It is clear now that the smartest strategy is the one in which we divide all possible events into two groups of equal probability.

It should be stressed that the value of the SMI is fixed once the game is fully described in terms of the distribution. It does not depend on the way we ask questions, or on whether we ask questions at all. The information is *there* in the very description of the game (or, more generally, in the specification of the rv). The number of questions and the amount of information obtained by each question can vary, but the total amount of information is fixed.

Notes on Chapter 3

Note 1 (solution to Exercise 3.5)

For notational simplicity, we use the notation X, Y, Z for X_1, X_2, X_3, respectively.

We have three experiments, or rv's $X \to Y \to Z$ forming a Markov chain. We want to prove that

$$I(X;Z|Y) = 0. \tag{1}$$

Note that by definition of the MI we have the inequality

$$I(X;Z|Y) \geq 0. \tag{2}$$

To prove the equality, we use the Markovian property in the form

$$p(z|y,x) = p(z|y). \tag{3}$$

Here, we can say that the probability of the outcome of the "present" experiment depends only on the recent "past" and not on the distant "past."

Using the definition of the conditional probability and the Markovian property (3), we write

$$p(x,z|y) = \frac{p(x,y,z)}{p(y)} = \frac{p(z|y,x)p(y,x)}{p(y)}$$
$$= p(z|y)p(x|y). \tag{4}$$

From the definition of the MI, we write

$$I(X;Z|Y=y) = \sum_x \sum_z p(x,z|y) \log\left[\frac{p(x,z|y)}{p(x|y)p(z|y)}\right]. \tag{5}$$

From Eqs. (4) and (5), it follows that

$$I(X;Z|Y=y) = 0. \tag{6}$$

Since this equality holds for any y [with $p(y) = p(Y=y) \neq 0$], it follows also that the average over all possible values of y is

$$I(X;Z|Y) = \sum_y I(X;Z|Y=y)p(y) = 0. \tag{7}$$

Thus, given the "present" Y, the "past" X, and the "future" Z become independent.

Note 2 [see Han and Kobayashi (2001)]

Note 3 (the general case: $0 < \alpha + \beta < 2$)

We prove here that the transition matrix $\boldsymbol{P}^n$ tends to a constant matrix when $n \to \infty$. We have excluded the equality $\alpha + \beta = 0$ in case (a), and the equality $\alpha + \beta = 2$ in case (b). We want to explore how $\boldsymbol{P}^n$ changes for large n.

We define the matrix

$$\boldsymbol{\Delta} = \boldsymbol{P} - \boldsymbol{I} = \begin{pmatrix} 1-\alpha & \alpha \\ \beta & 1-\beta \end{pmatrix} - \begin{pmatrix} 1 & 0 \\ 0 & 1 \end{pmatrix} = \begin{pmatrix} -\alpha & \alpha \\ \beta & -\beta \end{pmatrix}, \tag{1}$$

where $\boldsymbol{I}$ is the 2×2 unit matrix. Thus, we have

$$\boldsymbol{P} = \boldsymbol{I} + \boldsymbol{\Delta} = \boldsymbol{I} + \begin{pmatrix} -\alpha & \alpha \\ \beta & -\beta \end{pmatrix}. \tag{2}$$

We use the binomial theorem to sum on the right hand side of Eq. (2).

$$\boldsymbol{P}^n = (\boldsymbol{I} + \boldsymbol{\Delta})^n = \sum_{k=0}^{n} \binom{n}{k} \boldsymbol{I}^{n-k} \boldsymbol{\Delta}^k. \tag{3}$$

Note that we use here the binomial theorem for a sum of matrices. This is valid because the two matrices $\boldsymbol{I}$ and $\boldsymbol{\Delta}$ are commutative.

Clearly, the matrix $\boldsymbol{\Delta}$ is singular, i.e. its determinant is zero:

$$|\boldsymbol{\Delta}| = \begin{vmatrix} -\alpha & \alpha \\ \beta & -\beta \end{vmatrix} = \alpha\beta - \alpha\beta = 0. \tag{4}$$

The characteristic polynomial is

$$\begin{aligned} |\boldsymbol{\Delta} - t\boldsymbol{I}| &= \begin{vmatrix} -\alpha - t & \alpha \\ \beta & -\beta - t \end{vmatrix} = (\alpha + t)(\beta + t) - \alpha\beta \\ &= t^2 + (\alpha + \beta)t = 0. \end{aligned} \tag{5}$$

This polynomial has two roots:

$$t_1 = 0, \quad t_2 = -(\alpha + \beta). \tag{6}$$

The square of $\boldsymbol{\Delta}$ is

$$\boldsymbol{\Delta}^2 = \boldsymbol{\Delta}\boldsymbol{\Delta} = \begin{pmatrix} -\alpha & \alpha \\ \beta & -\beta \end{pmatrix} \begin{pmatrix} -\alpha & \alpha \\ \beta & -\beta \end{pmatrix} = -(\alpha + \beta)\boldsymbol{\Delta}. \tag{7}$$

Note that the last result may be obtained directly from the Cayley–Hamilton theorem, and that every matrix satisfies its characteristic equation, i.e. Eq. (7) can be obtained by substituting $\boldsymbol{\Delta}$ for t in the characteristic Eq. (5).

By repeating the multiplication in Eq. (7), we can obtain the nth power of the matrix $\boldsymbol{\Delta}$:

$$\boldsymbol{\Delta}^n = [-(\alpha + \beta)]^{(n-1)}\boldsymbol{\Delta}. \tag{8}$$

Substituting this result in the expansion of $\boldsymbol{P}^n$ in Eq. (3), we get (note that $\boldsymbol{I}^k = \boldsymbol{I}$ for any k and $\Delta^0 = \boldsymbol{I}$)

$$\begin{aligned}\boldsymbol{P}^n &= \sum_{k=0}^{n} \binom{n}{k} \boldsymbol{I}^{(n-k)} \boldsymbol{\Delta}^k = \boldsymbol{I} + \sum_{k=1}^{n} \binom{n}{k} \boldsymbol{\Delta}^k \\ &= \boldsymbol{I} + \sum_{k=1}^{n} (-\alpha - \beta)^{k-1} \binom{n}{k} \boldsymbol{\Delta}^k \\ &= \boldsymbol{I} + \boldsymbol{\Delta} \sum_{k=1}^{n} (-\alpha - \beta)^{k-1} \binom{n}{k}. \end{aligned} \tag{9}$$

We now use the identity

$$\begin{aligned}(1 - \alpha - \beta)^n &= \sum_{k=0}^{n} (-\alpha - \beta)^k \binom{n}{k} \\ &= 1 + \frac{1}{-\alpha - \beta} \sum_{k=1}^{n} (-\alpha - \beta)^{k-1} \binom{n}{k} \end{aligned} \tag{10}$$

and rewrite the right hand side of Eq. (9) as

$$\boldsymbol{P}^n = \boldsymbol{I} + \frac{(1 - \alpha - \beta)^n - 1}{-\alpha - \beta} \boldsymbol{\Delta}. \tag{11}$$

Since we assume that $0 < \alpha + \beta < 2$, it follows that (from $\alpha + \beta > 0$)

$$\alpha + \beta - 1 > 1 \tag{12}$$

or, equivalently [multiplying (12) by -1],

$$1 - \alpha - \beta < 1 \tag{13}$$

and similarly (from $\alpha + \beta < 2$)

$$\alpha + \beta - 1 < 1 \tag{14}$$

or, equivalently [multiplying (14) by -1],

$$1 - \alpha - \beta > -1. \tag{15}$$

The two results (13) and (15) can be written as

$$-1 < (1 - \alpha - \beta) < 1. \tag{16}$$

Thus, for large n, the quantity $(1 - \alpha - \beta)^n$ tends to zero. Therefore, we can write for $\boldsymbol{P}^n$ in Eq. (11), for large n,

$$\begin{aligned} \boldsymbol{P}^n \to \boldsymbol{I} + \frac{1}{\alpha+\beta}\boldsymbol{\Delta} &= \begin{pmatrix} 1 & 0 \\ 0 & 1 \end{pmatrix} + \frac{1}{\alpha+\beta}\begin{pmatrix} -\alpha & \alpha \\ \beta & -\beta \end{pmatrix} \\ &= \begin{pmatrix} 1 - \dfrac{\alpha}{\alpha+\beta} & \dfrac{\alpha}{\alpha+\beta} \\ \dfrac{\beta}{\alpha+\beta} & 1 - \dfrac{\beta}{\alpha+\beta} \end{pmatrix} \\ &= \frac{1}{\alpha+\beta}\begin{pmatrix} \beta & \alpha \\ \beta & \alpha \end{pmatrix}. \end{aligned} \tag{17}$$

This is an important result. We started with any stochastic matrix of the form (3.106) with the condition $0 < \alpha + \beta < 2$ and we found that $\boldsymbol{P}^n$ tends to a limit (17).

This means that, starting with any initial vector $\boldsymbol{\pi}^{(0)}$, after a large number of steps n, we get

$$\boldsymbol{\pi}^{(0)}\boldsymbol{P}^n \to \frac{\left(\pi_1^{(0)}, \pi_2^{(0)}\right)}{(\alpha+\beta)}\begin{pmatrix} \beta & \alpha \\ \beta & \alpha \end{pmatrix} = \frac{(\beta, \alpha)}{\alpha+\beta}.$$

Note 4

For the initial vector $\pi^{(0)} = (0.5, 0.5)$ and a stochastic matrix which depends on n,

$$\mathbf{P}(n) = \begin{pmatrix} 1 - \alpha(n) & \alpha(n) \\ \beta(n) & 1 - \beta(n) \end{pmatrix},$$

with $\alpha(n)$ and $\beta(n)$ as given in Exercise 3.7, and we find the results shown in Figs. N.1 and N.2.

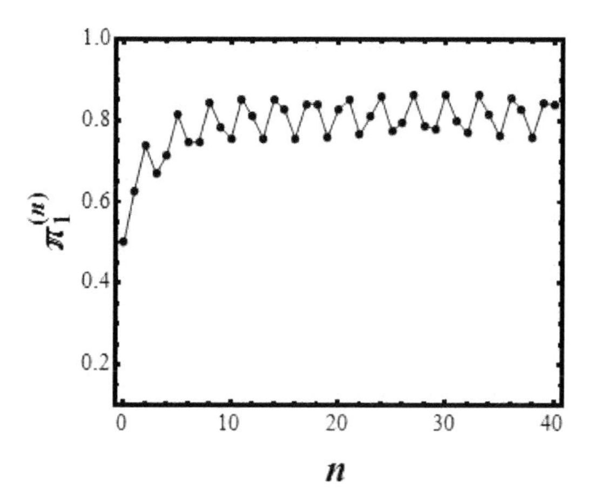

Fig. N.1. The variation of $p = \pi_1^{(n)}$ with n for exercise 3.7.

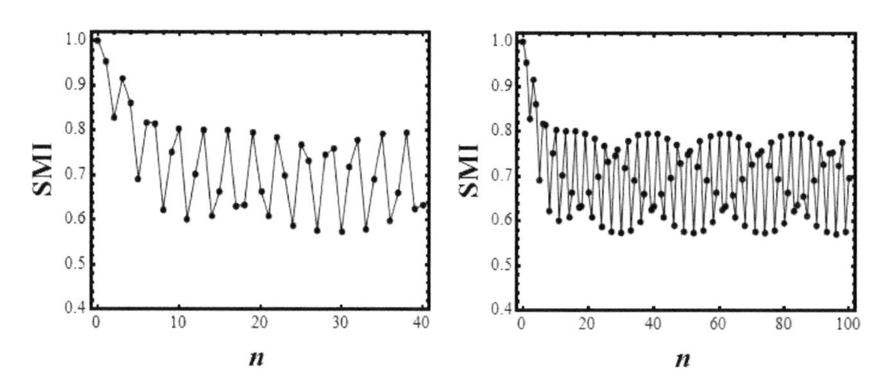

Fig. N.2. Variation of the SMI for $n = 20$ and $n = 1000$ as a function of n, for exercise 3.7.

Notes on Chapter 5

Note 1

Note here that we have said that the entropy of a well-defined isolated system at equilibrium does not change with time. The literature is replete with statements such as "Entropy tends to increase" or "Entropy tends to a maximum." Such statements are not only untrue — they are meaningless. One cannot say anything about the entropy changes without specifying the process and the system for which the entropy is defined.

Note 2

In an earlier book [Ben-Naim, (2012)], I have erroneously interpreted $\log h$ as an MI to account for the correlation between the location and the momentum. In fact, h measures the limit of the accuracy which is imposed on the product of $\Delta x \Delta p$, and is not a measure of the correlation between the location and the momentum.

Note on Appendix D

Note 1

Note that δ is used for the Kronecker delta and the Dirac delta, and for the functional derivative.

References and Suggested Reading

Ash, R. (1965), *Information Theory*. Interscience Publishers, John Wiley, New York.

Bell, J. E. (2003), *Evaluating Psychological Information*, 4th edition, Pearson.

Ben-Naim, A. (1987), Is mixing a thermodynamic process? *Am. J. Phys.* **55**, 725.

Ben-Naim, A. (1992), *Statistical Thermodynamics for Chemists and Biochemists*, Plenum, New York.

Ben-Naim, A. (2001), *Cooperativity and Regulation in Biochemical Processes*. Plenum Press, New York.

Ben-Naim, A. (2006a), *A Molecular Theory of Solutions*. Oxford University Press, Oxford.

Ben-Naim, A. (2006b), *Am. J. Phys.* **74**, 1126.

Ben-Naim, A. (2007), *Entropy Demystified: The Second Law of Thermodynamics Reduced to Plain Common Sense*. World Scientific, Singapore.

Ben-Naim, A. (2008), *A Farewell to Entropy: Statistical Thermodynamics Based on Information*. World Scientific, Singapore.

Ben-Naim, A. (2009), An informational-theoretical formulation of the second law of thermodynamics. *J. Chem. Educ.* **86**, 99.

Ben-Naim, A. (2010), *Discover Entropy and the Second Law of Thermodynamics: A Playful Way of Discovering a Law of Nature*. World Scientific, Singapore.

Ben-Naim, (2011a), *Molecular Theory of Water and Aqueous Solutions. Part II: The Role of Water in Protein Folding, Self-Assembly and Molecular Recognition*. World Scientific, Singapore.

Ben-Naim, A. (2011b), Entropy: order or information. *J. Chem. Educ.* **88**, 594.

Ben-Naim, A. (2012), *Entropy and the Second Law: Interpretation and Misss-Interpretations*. World Scientific, Singapore.

Ben-Naim, A. (2013), *The Protein Folding Problem and Its Solutions*. World Scientific, Singapore.

Ben-Naim, A. (2014), *Statistical Thermodynamics, with Applications to Life Sciences*. World Scientific, Singapore.

Ben-Naim, A. (2015a), *Information, Entropy, Life and the Universe: What We Know and What We Do Not Know*. World Scientific, Singapore.

Ben-Naim, A. (2015b), *Discover Probability: How to Use It, How to Avoid Misusing It, and How It Affects Every Aspect of Your Life*. World Scientific, Singapore.

Ben-Naim, A. (2016a), *The Briefest History of Time*. World Scientific, Singapore.

Ben-Naim, A. (2016b), *Myths and Verities in Protein Folding Theories*. World Scientific, Singapore.

Ben-Naim, A. (2016c), *Entropy, the Truth, the Whole Truth and Nothing but the Truth*. World Scientific, Singapore.

Bent, H. A. (1965), *The Second Law*. Oxford University Press, New York.

Boltzmann, L. (1877), Vienna Academy. **42**, "*Gesammelte Werke*" p. 193.

Boltzmann, L. (1896), *Lectures on Gas Theory*. Translated by S. G. Brush, Dover, New York (1995).

Brillouin, L. (1962), *Science and Information Theory*. Academy Press, New York.

Brush, S. G. (1976), *The Kind of Motion We Call Heat: A History of the Kinetic Theory of Gases in the 19th Century, Book 2: Statistical Physics and Irreversible Processes*. North-Holland.

Brush, S. G. (1983), *Statistical Physics and the Atomic Theory of Matter, from Boyle and Newton to Landau and Onsager*. Princeton University Press, Princeton.

Callen, H. B. (1960), *Thermodynamics*. John Wiley and Sons, New York.

Callen, H. B. (1985), *Thermodynamics and an Introduction to Thermostatics*, 2nd edition. Wiley, New York.

Cooper, L. N. (1968), *An Introduction to the Meaning and Structure of Physics*. Harper & Row, New York.

Cover, T. M. and Thomas, J. A. (1991), *Elements of Information Theory*. John Wiley and Sons, New York.

Denbigh, K. (1981), How subjective is entropy? *Chem. Brit.* **17**, 168.

Denbigh, K. G. and Denbigh, J. S. (1985), *Entropy in Relation to Incomplete Knowledge*. Cambridge University Press, Cambridge.

Denbigh, K. G. (1989), Note on entropy, disorder and disorganization. *Brit. J. Phil. Sci.* **40**, 323.

Dugdale, J. S. (1996), *Entropy and Its Physical Meaning*. Taylor and Francis, London; entropysite.oxy.com.

Eddington, Sir Arthur (1928), *The Nature of the Physical World*. Cambridge University Press.

Falk, R. (1979), *Understanding Probability and Statistics, A Book of Problems*. Wellesley, MA: A K Peters.

Fano, R. M. (1961), *Transmission of Information: A Statistical Theory of communication*. MIT Press, Cambridge, USA.

Fast, J. D. (1962), *Entropy: The Significance of the Concept of Entropy and Its Applications in Science and Technology*. Philips Technical Library, The Netherlands.

Gibbs, J. W. (1906), *Collected Scientific Papers of J. Willard Gibbs*. Longmans Green, New York.

Han, T. S. and Kobayashi, K. (2001), *Mathematics of Information and Coding: Translations of Mathematical Monographs*. American Mathematical Society, Rhode Island.

Hofstadter, D. R. (1985) *Metamagical Themas: Questing for the Essence of Mind and Pattern*. Bantam Books, New York.

Jaynes, E. T. (1957a), *Phys. Rev.* **106**, 620.

Jaynes, E. T. (1957b), *Phys. Rev.* **108**, 171.

Jaynes, E. T. (1965), Gibbs vs Boltzmann entropies. *Am. J. Phys.* **33**, 391.

Katz, A. (1967), *Principles of Statistical Mechanics: The Informational Theory Approach*, W. H. Freeman, London.

Khinchin, A. I. (1957), *Mathematical Foundation of Information Theory*. Dover, New York.

Lemons, D. S. (2013), *A Student's Guide to Entropy*. Cambridge University Press, Cambridge.

Lewis, G. N. (1930), The symmetry of time in physics. *Science* **71**, 569.

Lindley, D. V. (1965), *Introduction to Probability and Statistics*. Cambridge University Press, Cambridge.

Matsuda, H. (2000), *Phys. Rev. E* **62**, 3096.

McGill, W. J. (1954), Multivariate information transmission. *Psychometrika* **19**, 97.

Nordholm, S. (1997), In defense of thermodynamics — an intimate analogy. *J. Chem. Educ.* **74**, 273.

Patel, A. D. (2008), *Towards Understanding of the Orign of Genetic Languages, in Quantum Aspects of Life*, ed. by D. Abbott, P. C. W. Davies, and A. K., Pati. Imperial College Press, World Scientific.

Rawlinson, G. E. (1976), "The Significance of Letter Position in Word Recognition," Ph.D. thesis, Psychology Department, University of Nottingham.

Sackur, O. (1911), *Annalen der Physik* **36**, 958.

Shannon, C. E. (1948), A mathematical theory of communication. *Bell System Tech. J.* **27**.

Shannon, C. E. and Weaver, W. (1949), *The Mathematical Theory of Communication*. University of Illinois Press.

Tetrode, H. (1912), *Annalen der Physik* **38**, 434.

Thomson, W. (1874), *Proc. R. Soc. Edinb.* **8**, 325.

Tribus, M. and McIrvine, E. C. (1971), Entropy and information. *Sci. Am.* **225**, 179.

Velan, H. and Frost, R. (2007), Cambridge University versus Hebrew University; the impact of letter transposition on reading English and Hebrew. *Psychon Bull Rev* **14**: 913.

Yaglom, A. M. and Yaglom, I. M. (1983), *Probability and Information*. Translated from Russian by V. K. Jain, D. Reidel, Boston.

Index

Other Recent Books by the Author

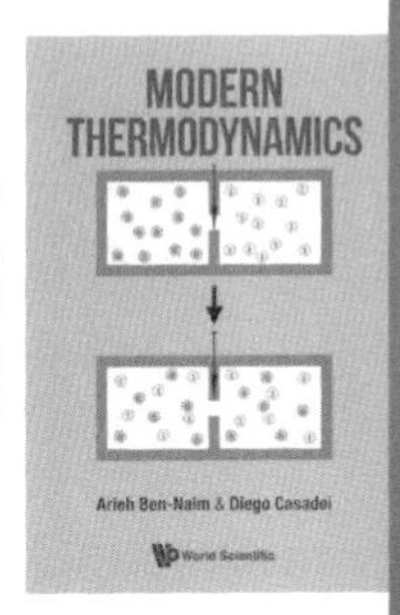

Modern Thermodynamics

By: Arieh Ben-Naim & Diego Casadei

ISBN: 978-981-3200-75-3
ISBN: 978-981-3200-76-0 (pbk)

Entropy
The Truth, the Whole Truth, and Nothing But the Truth

By: Arieh Ben-Naim

ISBN: 978-981-3147-66-9
ISBN: 978-981-3147-67-6 (pbk)

The Briefest History of Time
The History of Histories of Time and the Misconstrued Association between Entropy and Time

By: Arieh Ben-Naim

ISBN: 978-981-4749-84-8
ISBN: 978-981-4749-85-5 (pbk)

Entropy Demystified
The Second Law Reduced to Plain Common Sense
2nd Edition

By: Arieh Ben-Naim

ISBN: 978-981-3100-11-4
ISBN: 978-981-3100-12-1 (pbk)

Myths and Verities in Protein Folding Theories

By: Arieh Ben-Naim

ISBN: 978-981-4725-98-9
ISBN: 978-981-4725-99-6 (pbk)

Information, Entropy, Life and the Universe

What We Know and What We Do Not Know

By: Arieh Ben-Naim

ISBN: 978-981-4651-66-0
ISBN: 978-981-4651-67-7 (pbk)

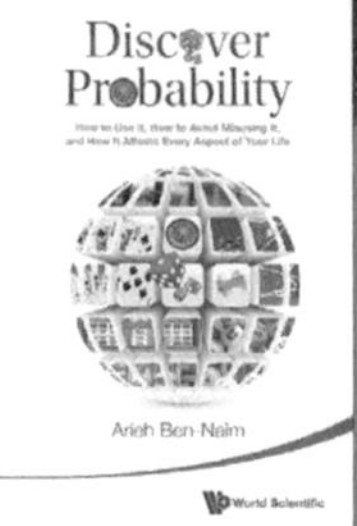

Discover Probability

How to Use It, How to Avoid Misusing It, and How It Affects Every Aspect of Your Life

By: Arieh Ben-Naim

ISBN: 978-981-4616-31-7
ISBN: 978-981-4616-32-4 (pbk)

Statistical Thermodynamics

With Applications to the Life Sciences

By: Arieh Ben-Naim

ISBN: 978-981-4579-15-5
ISBN: 978-981-4578-20-2 (pbk)

Alice's Adventures in Molecular Biology

Interpretation and Misss-Interpretationsss

By: Arieh Ben-Naim & Roberta Ben-Naim

ISBN: 978-981-4417-24-2
ISBN: 978-981-4417-25-9 (pbk)

The Protein Folding Problem and Its Solutions

By: Arieh Ben-Naim

ISBN: 978-981-4436-35-9
ISBN: 978-981-4436-36-6 (pbk)

Entropy and the Second Law

Interpretation and Misss-Interpretationsss

By: Arieh Ben-Naim

ISBN: 978-981-4407-55-7
ISBN: 978-981-4374-89-7 (pbk)

Alice's Adventures in Water-land

By: Arieh Ben-Naim & Roberta Ben-Naim

ISBN: 978-981-4338-96-7 (pbk)

Molecular Theory of Water and Aqueous Solutions

Part II: The Role of Water in Protein Folding, Self-Assembly and Molecular Recognition

By: Arieh Ben-Naim

ISBN: 978-981-4383-11-0 (Set)
ISBN: 978-981-4383-12-7 (pbk) (Set)
ISBN: 978-981-4350-53-2
ISBN: 978-981-4350-54-9 (pbk)

Discover Entropy and the Second Law of Thermodynamics

A Playful Way of Discovering a Law of Nature

By: Arieh Ben-Naim

ISBN: 978-981-4299-75-6
ISBN: 978-981-4299-76-3 (pbk)

Molecular Theory of Water and Aqueous Solutions

Part I: Understanding Water

By: Arieh Ben-Naim

ISBN: 978-981-283-760-8
ISBN: 978-981-4327-71-8 (pbk)

Printed in Great Britain
by Amazon